T0320296

Qualitative Modeling of Offshore Outsourcing Risks in Supply Chain Management and Logistics

What are the biggest challenges facing those managing supply chains? *Qualitative Modeling of Offshore Outsourcing Risks in Supply Chain Management and Logistics* is intended to benefit the stakeholders in client organizations by raising their understanding and awareness about the most dominant risks. This will equip supply chain managers to give more emphasis to mitigating these risks. It further showcases the development and validation of a conceptual framework that depicts the relationship among key offshore outsourcing risks. The text explores modeling various risks which disrupt the automotive supply chain and cybersecurity breaches in digital supply chains.

This book:

- Covers structural modeling of key offshore outsourcing risks for understanding their driving and dependence power.
- Presents a conceptual framework and hierarchical structural model for perfect order fulfilment in both upstream and downstream supply chains.
- Explores the challenges in handling operational risks associated with poor delivery performance or service quality.
- Models dimensions which affect vendor selection in offshore outsourcing environment.
- Investigates cultural influences on the management of geographically distributed operations in offshore outsourcing.
- Addresses the workforce-related offshore outsourcing risk such as loss of key professionals.

- Discusses the risk associated with selection of location, viz. distribution centres/warehouses in supply chain and logistics.
- Models dimensions related to cybersecurity breaches in digital supply chains because of IT offshoring.

It is aimed at senior undergraduate and graduate students, and academic researchers in the fields of manufacturing engineering, industrial engineering, mechanical engineering, supply chain management and production engineering.

Qualitative Modeling of Offshore Outsourcing Risks in Supply Chain Management and Logistics

Rajiv Kumar Sharma

CRC Press
Taylor & Francis Group
Boca Raton London New York

CRC Press is an imprint of the
Taylor & Francis Group, an **informa** business

Front cover image: dizain/Shutterstock

First edition published 2024
by CRC Press
2385 NW Executive Center Drive, Suite 320, Boca Raton FL 33431

and by CRC Press
4 Park Square, Milton Park, Abingdon, Oxon, OX14 4RN

CRC Press is an imprint of Taylor & Francis Group, LLC

ISBN: 978-1-032-46057-4 (hbk)
ISBN: 978-1-032-70391-6 (pbk)
ISBN: 978-1-032-70788-4 (ebk)

DOI: 10.1201/9781032707884

Typeset in Sabon
by codeMantra

Contents

Preface

Offshore outsourcing is a relatively new and demanding research field in international business operations. It is one of the ways through which the organizations try to address the new requirements of the marketplace. The strong theoretical development through qualitative models will accelerate the debate among the supply chain practitioners on the pertinence of offshore outsourcing in academics and industry.

According to a recent report by Grand View Research, the global outsourcing market was valued at USD 245.9 billion in 2021 and is projected to expand at a compound annual growth rate (CAGR) of 9.1% from 2022 to 2030.

Offshore outsourcing has implications for strategic management as well. There are important managerial implications of outsourcing for managers. It is important to consider to whom to outsource to (client aspect) and when to do so (time aspect). Trustworthiness of service provider(s), i.e., vendor(s), is also a matter of concern that leads to long-term difficulties with respect to various risks that are involved in offshoring business. The decision whether to outsource an activity or not is to be considered carefully and evaluated thoroughly in light of the background of the organization's business strategy. Consequently, each decision on outsourcing should be well thought-out in terms of effectiveness, efficiency and risk.

The concept of offshore outsourcing has gained popularity in the developing countries due to the accessibility of resources, viz. cheap labour and raw materials (Tjader, Shang, and Vargas 2010). Today, offshore outsourcing has become a primary choice among the service as well as manufacturing organizations to uphold their profit margins. To enhance the adoption rate of offshore outsourcing, numerous researchers provided a set of enablers and common drivers. By focusing on these enablers or so-called drivers, offshore outsourcing initiatives undertaken by businesses can be accomplished efficiently (Lahiri and Kedia 2011; Gunjan Yadav et al., 2018). However, the exact relationship among these enablers or driving factors needs to be understood so as to mitigate the risk associated with offshore outsourcing. Hence, this becomes extremely crucial to help managers/practitioners to understand the driver-dependence dynamics among various factors which affect the offshore outsourcing initiatives.

According to Behrooz Abdi, Vice President at Motorola, Inc., "Outsourcing is not just about cost benefits from lower salaries; by outsourcing to a highly skilled, readily available labor force overseas, companies can also improve their ability to compete in their fast paced domestic and international markets" (Wilson, 2003).

Despite the huge success of offshore outsourcing, many issues critical to offshore outsourcing business which require investigation and immediate attention are as follows:

- *How cultures influence the management of geographically distributed operations?*
- *How to maintain and improve the quality of offshore operations?*
- *How to reduce risk associated with opportunistic behaviour of a service provider in offshore outsourcing?*
- *How to tackle geo-political environment for offshore operations which help to mitigate risk?*
- *How to handle operational risks associated with poor delivery performance or service quality?*
- *How to address the workforce-related offshore outsourcing risk, i.e., loss of key professionals?*
- *What are the various dimensions related to cybersecurity breaches in digital supply chains because of IT offshoring?*

The answers to these questions need to be addressed in order to ensure the success of offshore outsourcing. It has been observed that India should improve in the field of service sector and mitigate all risks related to service/client organizations, especially those risks which are involved in offshore outsourcing. No doubt that in the present times India is focusing on the manufacturing sector, but it should not lose the upper hand in the service sector which has been developed for so many years. To this effect, this book investigates various risks and their inter-dependency as well as factors of many important risks and also inter-dependency among those risk dimensions, which are concerns for offshore outsourcing in the context of supply chain. It will help the managers to know which factor of risk is more important in terms of influencing other dimensions, and it is reflected by enhancing that particular risk. There are key process areas with respect to offshore outsourcing business environment which are of interest to client/service organizations. For instance, some of these process areas are listed as under:

- *Selecting a service provider organization*
- *Selecting the nature of work to be outsourced*

- *Dealing with cultural differences*
- *Creating and managing the offshore contract*
- *Contract negotiation*
- *Monitoring the ongoing outsourcing contracts*
- *Layoff of the existing workforce*
- *Transitioning work at offshore*
- *Establishing and improving service-level metrics*
- *Adaptation with offshore outsourcing environment*
- *Managing costs associated with in-bound, out-bound and process logistics*
- *Managing risk associated with intellectual property rights*
- *Synchronizing activities among stakeholders and working in coordination to complete the project*
- *Managing cybersecurity breaches in digital supply chains*

In the last two decades, numerous approaches, empirical as well as conceptual models, have been developed by researchers to study or model the impact of various risks on offshore outsourcing. But very limited work is found on the association among various types of offshoring risks. Because of the complex nature of offshore outsourcing and its interface between cultures, organizations, disciplines, technologies and tacit knowledge of employees, it is not easy to analyse the inter-relationships among the offshoring risks. As a problem, the number of key risks such as risk due to culture differences, opportunistic behaviour risk, political risk, intellectual property risk, financial risk, compliance and regulatory risk, and organization structural risk, influences offshore outsourcing risk. This book will address the following issues in the form of book chapters.

This book will address various topics related to offshore outsourcing risks in the following chapters:

- Chapter on key offshore outsourcing risks with details on the key influencing dimensions of various offshore outsourcing risks
- Chapter on structural models of key offshore outsourcing risks for understanding their driving and dependence power
- Chapter on vendor selection in an outsourcing environment
- Chapter on dimensions responsible for perfect order fulfilment in the supply chain network
- Chapter on the risk associated with location of warehouse facility in offshore outsourcing
- Chapter on key human dimensions for mitigating lack of coordination risk in supply chain
- Chapter on modeling dimensions related to cybersecurity breaches in digital supply chains because of IT offshoring
- Chapter on modeling various risks which disrupt the automotive supply chain with an illustrative case study (Covid-19 pandemic).

Acknowledgements

This book is the outcome of my interest in the world of supply chain management and logistics spanning the last one decade. This journey has been influenced by my mentors, colleagues, students and industry practitioners. My mentors from Indian Institute of Technology (IIT) Roorkee, specifically Prof. Pradeep Kumar and Prof. Dinesh Kumar, have influenced the way I look at and perceive the world of supply chain and logistics management. My colleagues and co-researchers have contributed significantly to the various ideas and frameworks that I have developed in the field of offshore outsourcing in supply chain and logistics.

Several ideas in this book with respect to offshore outsourcing risks have been supported by the respondents working as technology service providers, financial service providers, analytics service providers, business process service providers and providers of research and development solutions. Also, the services of various experts from the academia and industry are acknowledged.

I am grateful to my postgraduate and doctoral students for their support. A special thanks to Dr. Prashant Chauhan, Dr. Pratima Mishra, Rahul, Vishal and Mudit Rawat for their contributions. The outcome of my interactions with these students and industry participants has found a place in the discussion on various aspects of outsourcing decisions in supply chain and logistics.

I also wish to thank Gauravjeet Singh Reen, Senior Commissioning Editor-Engineering, CRC Press, Taylor & Francis Group for his enthusiasm, patience and support since the beginning of this book, as well as Aditi Mittal, Isha Ahuja and Mehnaz Hussain, Editorial Assistant-Engineering CRC Press, Taylor & Francis Group and Uma Maheswari, Project manager -PMO codeMantra their staff for their meticulous effort regarding production.

I am indebted to persons from the organization who directly or indirectly helped during the preparation of this book. It is my pleasure to acknowledge their help. I would like to thank my parents and family members for their encouragement.

Rajiv Kumar Sharma

About the author

Dr. Rajiv Kumar Sharma is presently working as an Associate Professor in the Mechanical Engineering Department at NIT Hamirpur. He has also served as the Head of the Department of Mechanical Engineering. He has done his Bachelor's degree in Mechanical Engineering (with Hons) and Master's degree in Industrial and Production Engineering with Gold Medal from Thapar University, Patiala. Thereafter, he obtained his doctoral degree (PhD) from the IIT Roorkee, Uttarakhand. Till now, he has guided 6 PhD thesess, 28 MTech theses and more than 70 UG projects in various fields of engineering. He has research and teaching experience of more than 20 years and about 100+ publications in international journals of repute such as *International Journal of Production Research, International Journal of Quality & Reliability Management, Industrial Management & Data Systems, International Journal of Systems Science, Journal of Loss Prevention in the Process Industries, Reliability Engineering & System Safety, Quality and Reliability Engineering International, International Journal of Manufacturing Technology and Management, Total Quality Management & Business Excellence and Proceedings of the Institution of Mechanical Engineers, Part B & C: Journal of Engineering Manufacture* published by well-known academic publishers such as Elsevier, Taylor & Francis, Wiley, Emerald, Sage and Inder-science. He has delivered expert talks in faculty development programmes, short-term training programmes under the aegis of AICTE ATAL ACADEMY organized by NIT Warangal, NIT Calicut, MNIT Jaipur, NIT Jalandhar, NIT Hamirpur and SVNIT Surat. He is also on the Editorial Board of many reputed journals such as *International Journal of Quality and Reliability Management, TQM Journal,* Emerald Publishers; *International Journal of Strategic Business Alliance,* Inderscience Publishers; and *Advances in Production Engineering & Management,* Production Engineering Institute (PEI), University of Maribor, Maribor, Slovenia European Union (EU). He has authored the book *Quality Management Practices in MSME Sectors for Practicing Engineers and*

Research Scientists published by Springer. He has completed a sponsored project from UGC and DST. His research interests include supply chain and logistics, flexible manufacturing systems, statistical quality control system reliability and maintenance, quality engineering, Six Sigma, lean manufacturing and Industry 4.0.

Chapter 1

Offshore outsourcing, its types, benefits and risk categories

1.1 INTRODUCTION

Outsourcing is a business agreement in which firms, either national and/ or international, make a contract to carry out internal and/or external functions which may be value-added or non-value-added functions to experienced supplier(s) with an aim to gain a competitive advantage (Di Mauro et al. 2018; Uygun et al. 2022). Outsourcing is also defined as the business practice of hiring an outside party to perform the services or to make products. The term outsourcing entered the business dictionary in the 1980s. In the mid-20th century, as businesses grew, more specialized skills were required; companies perceived that external parties complete the work faster and more efficiently. This led to signing of contracts with external service providers to manage projects and business functions. With the developments in telecommunications and transportation infrastructure in the end of the 20th century, it became even more proficient to accomplish the work at distributed locations, mostly in developing countries like India, China and Brazil where salary packages are less as compared to that in the developed countries.

The biggest advantage of offshore outsourcing is time and cost savings. For example, a company making personal computers might trade internal parts for its machines from other vendors to minimize production costs. Without investing large amounts of money on the technology infrastructure, a law company might store and back up its files using a cloud service provider. In present times, the global outsourcing market is dominated by information technology (IT) outsourcing and business process outsourcing (BPO) as two sub-industries. IT outsourcing covers cloud computing services, cybersecurity, web hosting and data backups. BPO covers functions ranging from human resource services to customer services including marketing and logistics business functions. The outsourcing statistics (Deloitte Global Outsourcing Survey 2020, ISG Momentum® Market Trends & Insights Geography Report 2019; business services, clutch report small business outsourcing statistics in 2019; Outsourcing of payroll by

DOI: 10.1201/9781032707884-1

organization size worldwide by size 2019 Published by Statista Research Department, Jul 6, 2022) are as follows:

- *Global expenditure on outsourcing could hit $731 billion in the year 2023*
- *37% of small business companies subcontract at least one business process*
- *9% of the Philippines gross domestic product (GDP) comes from BPO*
- *70% of British B2B companies outsource key business operations*
- *92% of G2000 companies use IT outsourcing*
- *Over 1 million employees join every year in the outsourcing industry in China.*

1.2 OFFSHORE OUTSOURCING

As discussed in the introduction section, in outsourcing, one company provides services to another company. Offshore literally means any country other than your own country. When the outsourcing process is combined with the offshoring process, work is not only subcontracted out but is also performed in a different country in order to avail the benefits of both outsourcing and offshoring. Since the mid-1980s, outsourcing has become a prominent practice in manufacturing companies, which allows them to concentrate on core competencies (Rahman et al. 2020). Evolution of offshore outsourcing with four maturity stages, i.e., embryonic, growth, mature and ageing, is presented in Figure 1.1.

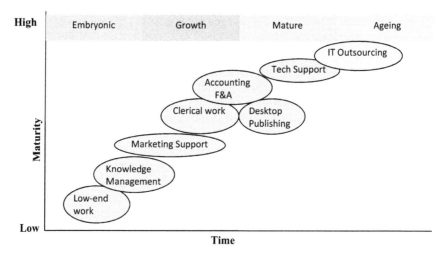

Figure 1.1 Evolution of offshore outsourcing for corporations.

Source: Aggarwal (2011).

In the present-day global setup, offshore outsourcing is unavoidable as it helps to promote coordination among supply chain members under which a third-party service provider is hired to perform certain business functions in an overseas country. For instance, automobile companies took outsourcing as a means to collaborate with competitors and developed shared solutions to accomplish green supply chain challenges. With advancement in communication infrastructure and easily available workforce at the global level, today organizations reap benefits of offshore outsourcing and view these relations as a "win–win" situation and not a zero-sum game. The key services under offshore outsourcing include contracts related to manufacturing, distribution and technological collaboration. With the increase in globalization and technological advancements, today businesses strive to expand their product development capabilities as well as production and volume flexibilities by taking into account the outsourcing services provided by third parties. According to Lockamy and McCormack (2010), today firms are progressively taking outsourcing options as a key driver of their diverse supply chain policies to attain competitive positioning via reducing costs and improving responsiveness. In the present-day service-oriented economy, do-or-buy decisions have taken over traditional make-or-buy decisions, which reveals a strategic question, i.e., whether services should be contracted to external entities or not? (Rahman et al. 2020).

In real terms, in offshore outsourcing, more risks are involved as compared to domestic outsourcing because of political and economic uncertainties, cultural differences, intellectual property and knowledge transfer risks. Offshoring is an inherently risky business because of the complexities involved in achieving management oversight and overseas control. Kaur, Singh, and Majumdar (2019) developed a model to address offshoring and outsourcing decisions to facilitate the decision-makers. The article by Bruccoleri et al. (2019) suggested that offshore outsourcing reduces the product recall magnitude as compared to captive outsourcing which increases it. Some of the views and definitions related to offshore outsourcing practice are listed in Table 1.1. The various steps involved in offshore outsourcing are discussed as under:

1. *Project/process selection*: This step includes project/process selection to be assigned to the offshore development team. The feasibility, benefits, challenges and risks are analysed before finally contracting a particular process or a project to the offshore company.
2. *Development of implementation plan*: In this step, a roadmap is developed which steers the way the client and Partner Company moves for detailed plan with focus on the project deliverables.
3. *Selection of vendor company*: This is a very crucial step as it deals with the selection of right outsourcing company which is important for the project to be successful.

Table 1.1 Views and definitions about offshore outsourcing

Author(s) & year	Views and definitions
Raiborn et al. (2009)	In the present-day service-oriented economy do-or-buy decisions have taken over traditional make-or-buy decisions that reveal about the strategic question, i.e., whether services should be contracted to external entities or not?
Hahn et al. (2011)	Offshore outsourcing is an industry term which indicates that work has been allocated overseas and is going to be performed by a third-party service provider in place of being carried out by the owner.
Mihalache et al. (2012)	Offshore outsource refers to the shifting of business operations to overseas countries to take the advantages of time and cost savings.
Bhattacharya et al. (2013)	The relationships may be viewed as a "win–win" situation and not a zero-sum game where the business owners plan to offshore their business functions to third-party service providers.
Luthra et al. (2014)	Companies working in the automobile sector used offshore outsourcing as a means to collaborate and work together with market leaders to meet the challenges.
Johnson and Graman (2015)	Recognized as a business strategy in 1989, outsourcing became an integral part of business economics in the 1990s. Often, outsourcing enables the businesses to focus on its core operations.
König and Spinler (2016)	A conceptual risk management framework is developed and the effect of logistics outsourcing on the supply chain vulnerability of shippers is discussed.
Pisani and Ricart (2016)	Offshore outsourcing of services is defined as the relocation of business processes at the international level instead of performing them in the host country.
Munjal et al. (2018)	There is a positive influence of imported technology and expert services from offshore outsourcing on firm performance. Outsourcing of specialized resources enhances firm performance.
Ishizaka et al. (2019)	In the literature, two forms of offshoring are: (i) "international outsourcing" in which contract is given to a foreign vendor and (ii) a captive model, in which an activity is performed in a firm's own subsidiary.
Nordås (2020)	Author developed outsourcing measure by relating the service functions provided by outside suppliers with that executed by workers inside manufacturing firms.
Uygun et al. (2022)	Companies usually face a number of trade-offs during outsourcing. To perceive a positive effect of outsourcing balance consisting of hidden and transaction costs and savings resulting from innovation and workforce must be achieved.

4. *Formation of team*: In this step, a team is formed for the IT development of project with the help of professional third-party offshoring service provider.
5. *Development of joint plan for execution and communication thereof*: The process owner (client) and the vendor should come together to develop mutually agreed joint implementation plan with necessary documentation and controls. Also, a communication plan for effective communication should be established and communicated to the entire team.
6. *Project monitoring and controlling*: With focus on the project/process deliverables, budget limits, project monitoring and controlling, all such activities are being done at the level of service provider. Figure 1.2 presents visual schema of steps involved in offshore outsourcing.

1.3 TYPES OF OFFSHORE OUTSOURCING

1.3.1 Business process outsourcing

Gartner Dataquest describes BPO as the assignment IT demanding business operations to a third party, who in turn controls and manages certain business operations, agreed upon pre-defined and quantifiable performance measures. It can also be stated as service of contract in which one or more business processes are transferred to a third-party service provider, where the latter assumes the responsibility of working out management, support and infrastructure requirements for the completion of the project. When these business operations are being executed by overseas vendors, it is termed as offshoring or offshore outsourcing. It encompasses the real relocation of the physical manufacturing facilities in foreign countries, generally in terms of labour and materials costs savings. Normally, the offshore outsourcing processes can be grouped into front-office and back-office solutions. The front-office solutions comprise service support to customer through call centres, help desks and providing financial, telemarketing

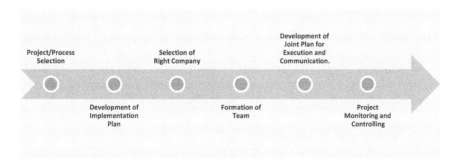

Figure 1.2 Procedural steps involved in offshore outsourcing.

(inbound/outbound) and technical services. The back-office solutions are related with processes like workforce and recruitment, accounting, mobile and web-app development, etc. India presently is the destination of choice for outsourcing because of various benefits, viz. large pool of skilled professionals, low-cost skilled and semi-skilled labour, English language–speaking workforce, high-quality work processes, scalability, and formidable support from central and state government agencies.

1.3.2 Infrastructure and technology outsourcing

Infrastructure and technology outsourcing is related to a business practice where the customer hands over some of its IT-related value chain processes which were done earlier in-house to outside agents called vendors. This involves services that provide necessary support to organizations, viz., network support services. IT outsourcing makes use of a third party to provide services rather than using in-house resources. Given the multitude of business operations and minute-by-minute task requirements, some of the more popular services under IT outsourcing are help desk and application development, data centre, desktop/personal computers and networking support services (local area networks (LANs), Wireles access networks (WANs)). In present times, it has become normal to outsource IT services to vendors. In the 1990s, the communication infrastructure helped India, with its large pool of English language speakers and technically qualified workforce, to attract firms such as Microsoft, Hewlett Packard Enterprise (HP), The International Business Machines Corporation (IBM), Oracle Corporation and Intel. IT outsourcing companies helped many of their clients working in various countries to implement business improvement solutions for managing their supply chains. According to a joint work undertaken by Klynveld Peat Marwick Goerdeler (KPMG) and HfS Research, the global market for IT outsourcing was valued at $648 billion in the year 2013 which is anticipated to rise at 4.7% annually until the year 2017 (HfS Research, 2013). It also stated that information technology is among the top three business activities being outsourced by the companies working in diverse business domains, such as manufacturing, energy, telecom, insurance, banking, gaming and retail (Mehta and Mehta 2017). According to a report published in Statista, the European IT outsourcing business is estimated to reach $103.9 billion by the year 2021, a growth trend which is observed universally. This growth comprises IT application outsourcing, IT infrastructure outsourcing and IT administration outsourcing (published by Statista Research Department, February 11, 2022).

1.3.3 Software outsourcing

Software outsourcing is a multidimensional and complex activity in which the prospective clients and vendor work together to produce and deliver the required software services. In software outsourcing, third-party services are taken to handle projects related to software development. These services vary from development of software to manage business practices or development and maintenance of existing software applications. According to a Statista report, software development is one of the most commonly outsourced IT functions. Though business organizations can do it on their own, it requires skilled developers and other resources to complete the project with time and cost constraints. Outsourced software projects contain substantial technical activities. It is also associated with a process of acquiring and integrating knowledge from key stakeholders, viz. customers, project managers and developers. For timely completion of projects, appropriate organizational controls are important to safeguard the interests of the stakeholders. India, China and Russia are the three countries which offer software development services across the globe. India is one country which can handle multifaceted software development projects with a promise to complete them in time. However, in the present times, high-tech product organizations with their offices in Silicon Valley are offshoring their software development work to countries such as Colombia, Mexico, South Africa, Belarus and Ukraine with an aim to access the highly skilled manpower in these countries which offers them significant cost savings and shorter cycle time.

1.4 BENEFITS OF OFFSHORE OUTSOURCING

1.4.1 Cost savings

Cost minimization is often a primary reason for offshore outsourcing, and the expectations are that significant savings will occur through outsourcing. The main reason why most of the companies prefer outsourcing of services through overseas service providers is the significant cost savings and timeliness of operations. Literature in the offshore outsourcing field reveals that outsourcing, on an average, results in cost savings of around 15%. These savings came in the form of cheaper labour, cheaper materials, greater efficiency and increased service offerings. An industrial labour report (India versus Germany) presented by Burger (2007) on successful offshore outsourcing projects mentions "significant cost saving" with timely delivery of good-quality software. Findings by Smite et al. (2015) and Nidthida Lin (2020) suggested that global outsourcing of complex project development tasks also finish successfully with significant cost savings and innovation performance.

1.4.2 Access to specialized expertise

In the specialization process, organizations outsource non-core activities to third parties for getting them completed in a cost-effective and timely manner. Thus, non-core activities for one organization become the core activities of another (Nordås, 2020). Luo et al. (2013) in their work emphasized upon the specific nature of offshore services, and their results showed that in these services the governance mode is a joint venture or partial ownership as these services are provided by specialized people. Such services include information technology or financial services which are supposed to be expensive and competitive in the business owners' country. Offshoring these specialized business functions, which is also recognized as knowledge process outsourcing (KPO), gives the business owners an opportunity to connect with human resources equipped with a specialized skill set at a lower cost, which not only results in lower hiring costs but also improvement in business results. The most common knowledge process works under KPO that are outsourced internationally are related to marketing, legal services, intellectual property rights (like for filing of patents, copyrights), offline and online training services, and activities related to research design and development.

1.4.3 Round-the-clock availability

Offshoring gives companies round-the-clock uptime because the employees work in geographically distributed time zones. Today's customers demand 24×7 services. The advent of internet has extended the normal business hours as nowadays when the client faces any problem with an offshore service provider, they wish to get it resolved instantly irrespective of the time. By capitalizing their workforce in different time zones, offshoring provides business houses a more realistic way to serve the customers round the clock with their support. Indian KPOs usually employ a team that works for USA clients depending upon the time when the US company functions or vice versa when American KPOs assign work to its Indian KPO partners; the time difference is about 12 hours. One can allocate work and take adequate rest as the allocated work is supposed to be completed by the time when one wakes up in the morning. Therefore, tasks go round the clock and such outsourcing of tasks work well in weekends also. Time zone differences offer the KPOs an opportunity to use the dedicated pool of in-house resources whenever they require them.

1.4.4 Focus on core competencies

Offshore outsourcing of services provides the business houses the flexibility to focus on their important activities and handles all the other development-related jobs. For instance, when a firm hires a developer for software development activities, the firm mainly focuses on its routine assignments, and the hired resource provider thus concentrates on the assigned outsourced work. This helps businesses to lessen the workload of their in-house pool of workers, thereby completing the projects within the assigned timeframe.

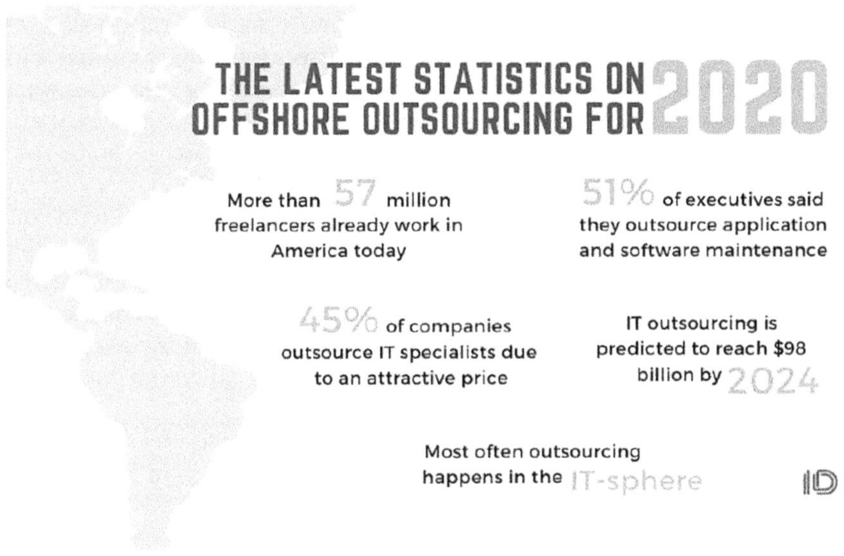

THE LATEST STATISTICS ON 2020
OFFSHORE OUTSOURCING FOR

More than 57 million
freelancers already work in
America today

51% of executives said
they outsource application
and software maintenance

45% of companies
outsource IT specialists due
to an attractive price

IT outsourcing is
predicted to reach $98
billion by 2024

Most often outsourcing
happens in the IT-sphere   ID

Figure 1.3 Statistical data on offshore outsourcing benefits.

Source: https://internetdevels.com/blog/offshore-outsourcing-aftermath-covid.

Thus, in addition to potential cost savings, the greatest advantage of offshoring services is to free up own resources that can be utilized to focus on core business functions, which finally make more profit margins.

For example, when your company outsources a data entry job, all the related requirements are managed by companies that are experts in handling data entry, scanning, processing and indexing your data. These are the benefits of BPO. Your data are always in good shape and accessible to you whenever you need them. Another example may be of the human resource department of a company which spends 4 hours/week in performing review and approval of worker timesheets. By offshoring this task to a knowledge-based offshore partner, the firm can spend more hours on those business activities that will generate revenue. Figure 1.3 presents the latest statistical data on offshore outsourcing benefits.

1.5 OFFSHORE OUTSOURCING RISK CATEGORIES

When outsourcing activity is accomplished through offshore vendors, it results in additional risks in terms of language barriers, differences in time zones and cultures, and geographical barriers. Risks also result from opportunistic behaviour of either of the parties, i.e., vendor and client.

This type of behaviour hinders cooperative innovation, and it may result in shirking of responsibilities among the parties. Some structural risk often arises when vendors stop training their employees who are entrusted with the job of negotiating with the clients. Another outsourcing risk arises when vendors make alterations to the business processes, technologies used and operating procedures without taking the clients into confidence. Some risks related to cost dimensions are also observed in outsourcing, viz. switching costs, unforeseen transitions and management costs, cost related with contract amendments, and cost to address disputes and litigations. According to Ishizaka et al. (2019), outsourcing business is associated with numerous risks which often lead to non-completion of projects in time. The fact that the workforce, the greatest resource of any concern, could leave is the risk associated with the outsourcing business, and this is called the risk of loss of core professionals.

On the basis of discussion presented in the section, the various risks are grouped into ten categories, as shown in Table 1.2 and also in Figure 1.4.

1.5.1 Political risk

Political risk arises when policies of a host country become unsupportive for delivery of projects and adversely affect organizations' profits (Hansen et al. 2018). It can also be defined as interference of external agents with or without governmental sanctions which originate either from within or outside the host country, and considerably impact the supplier's capability to ensure the delivery of products or services with time and cost constraints. Such type of risks include geo-political risks, sovereign risk, transactions risk, etc., and political risks are related with different provinces with totally dissimilar sociopolitical environments (Hansen et al. 2017, 2018). Political risk mainly involves confiscation of properties, denial of contract agreement, inequitable taxation norms, trade restrictions, expropriation of assets, nationalization or disturbance risk. The uncertainty in the political system in less developed countries is usually anticipated. It not only increases contract costs but also jeopardizes the project's completion. Results presented by Ancarani et al. (2015) show that dimensions such as the host country, domestic country, nature of the industry and size of the business firm considerably affect the time period of offshoring work. The licence and contract agreement and foreign direct investment options for clients mainly depend on international business theory which includes the risks associated with expropriation of assets of companies, i.e., this risk arises when government forcibly takes over the possession of property without proper compensation. In emerging economies, risks due to uncertain economic and political environment are more, because of insufficient government support. Occasionally, relations between countries substantially impact the business environment between the host and outsourcing

Table 1.2 Definition of offshore outsourcing risks and their key determinants

S. No.	Name of risk	Definition	Key determinants
1	Political risk	The interference of external agents with or without governmental sanctions which originates either within or outside the host country, and considerably impacts the supplier's capability to ensure the delivery of products or services with time and cost constraints.	Civil disturbance, financial and monetary policies, terrorism, industrial labour relations, domestic policies of country, bribery, global relationships, availability of human resources.
2	Cultural differences	Cultural difference is related to deep-seated attitudes and beliefs which are often more difficult to witness than differences due to language, so they may go unnoticed.	People's values, beliefs and attitudes; managing time zone differences, food habits, dress and appearance, sense of self; learning strategies, work habits, procedures and practices, conversation language.
3	Opportunistic behaviour risk	Opportunistic behaviour risk hinders cooperative innovation and it may result in shirking of responsibilities among the parties.	Communication gap, geographical distances, incomplete work specifications, level of control, post contractual behaviour, potential transaction costs, under-investment in project, switching cost.
4	Intellectual property infringement risk	In an outsourcing arrangement, the firm has to share its intellectual property with the prospective vendors. So, such risks are associated with transferring intellectual property to an outsourcing partner.	Information technology, opportunistic behaviour, lawless environment, technical knowhow, research and development process, project size.
5	Financial risk	Such risks are related to cost dimensions in outsourcing, viz. switching costs, unforeseen transitions and management costs, cost related with contract amendments, cost to address disputes and litigations.	Adaptation cost, disputes and litigations, costly contractual amendments, measurement problems, switching cost, service provider selection cost, layoff cost, transition and management cost.

(Continued)

Table 1.2 (Continued) Definition of offshore outsourcing risks and their key determinants

S. No.	Name of risk	Definition	Key determinants
6	Compliance and regulatory risk	Regulatory risk refers to the risk associated with change in the laws or regulations by the government or regulatory body that impacts a business contract among the parties.	Brand reputation, meeting demand of investors, fraud and security breaches, defence against legal repercussions, change in tax structure and data protection issues.
7	Organization structural risk	Organizational structural risk arises when vendors stop training their employees who are entrusted with the job of negotiating with the clients.	Regulatory issues, incompatibility, delay in completion of project, inappropriate resource allocation, deficiency in capabilities and reduction in human capital.
8	Operational risk	This risk arises when the third party to whom services are contracted fails to deliver them as per contract.	Poor delivery performance, lack of competency to fulfil task, process fragmentation, conflict of objectives.
9	Loss of core professionals	The risk arises with loss of core professionals or knowledge pool the organization has to handle the outsourcing activities.	Reduction in work force, loss of knowledge pool, layoff, lack of communication, exploitation of experts.
10	Cybersecurity risk	It is defined as "defences against electronic attacks launched via computer systems".	Cross-country risk, within-country risk, physical threats, software vulnerabilities, cyber assaults or attacks, insider threats because of human, and contextual risk factors.

Figure 1.4 Different types of offshore outsourcing risks.

companies. Political instability in a country affects offshore destinations and is thus considered as one of the distinct risks which contributes to offshore outsourcing. Hahn and Bunyaratavej (2010) stated that client organizations are averse to foreign countries where political risk is high, and they prefer to do business in countries with stable political environments. Infrastructural problems or political instability can affect the progress of work and thus contribute to failure. The political unrest in a country deters business dealings and refrains the firms from setting up operations in such nations. Political risks because of uncertain political environments also influence financial risks. Summarizing the literature studies, dimensions responsible for political risk are classified into eight categories, as shown in Figure 1.5.

The propagation of mass destruction weapons has created an insecurity that is moreover related to interstate and civil disturbances, and it further adds to the political risk (Chauhan et al. 2015; Hansen et al. 2017). Also, the risks that contribute to political risk include security exposure linked to politically driven riots, strikes, sabotage and cross-border terrorism, which impacts on operational performance of business firms in a direct or indirect manner. Political risk undesirably affects offshore outsourcing works, viz.

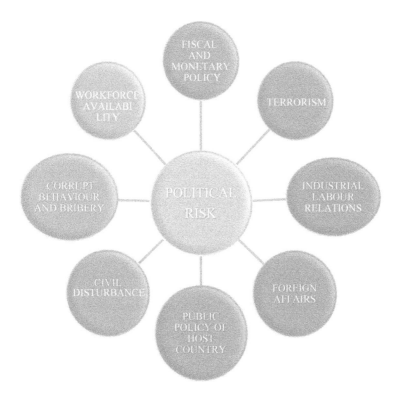

Figure 1.5 Categories of political risk.

BPO, Information Technology Outsourcing (ITO), and KPO services. Details of political risk dimensions along with definitions are depicted in Table 1.3.

1.5.2 Risk due to cultural differences

In the present era of globalization, the collaboration among countries is important. Culture plays an essential role as it limits the accessibility of the best workforce with necessary skill set. Additional risks are observed in outsourcing business environment when service providers work with distinct cultures, language differences, and topographical and time-related zone differences (Rahman et al. 2020). In order to evade the unsuccessful offshoring results, there should be flexibility among the professionals from various outsourcing destinations. Cultural difference is related to attitudes and beliefs which are difficult to witness than language differences, so they may go untraced. The offshore service providers handle the language and culture barriers with better collaboration with the clients. Rahman et al. (2021) reiterated that both vendors and clients get disturbed with cultural and language barriers being faced by teams before undertaking outsourcing decisions in order to ensure smooth functioning of offshoring services globally.

Table 1.3 List of dimensions of political risk with definitions

S. No.	Political risk dimensions	Definition
1	Civil disturbance	Use of weapons between the citizens of country during conflict is called civil disturbance–like situation.
2	Fiscal and monetary policy	Rules and regulations related to financial matter and valuation of the currency comes under financial and monetary policies.
3	Terrorism	The unauthorized/unofficial use of violence against the will of the country is called terrorist activity.
4	Industrial labour relations	Industrial labour relation means settling employee's problems with management through mutual cooperation and agreement.
5	Public policy of host country	Public policies are administrative decisions which are related to all activities and issues within a country.
6	Corrupt behaviour and bribery	Corruption is the misuse of power by the authorities and may be accepted by the system of the country.
7	Foreign affairs	The political relationship between two countries is defined as the foreign affairs.
8	Workforce availability	The non-availability of work force with the required skill set and experiences do limit the quality of offshoring service.

When the supply chain becomes disruptive, it may lead to "massive economic damage" for the client organizations. The different national cultures create instability in business operations which incurs additional transaction costs. Various researchers (Chang and de Búrca 2016) indicate cultural difference as the key stumbling block of IT offshore outsourcing. Cultural differences in service provider companies were considered as one of the major risk factors in offshoring services (Tate and Ellram 2012). When members of project handling teams are of different cultural backgrounds, it becomes an uphill task to have mutual understanding among all members of the team. Not only time and cultural difference but also communication gap is a major issue which is faced by both teams engaged in offshoring work (Karlsen et al. 2021). Various dimensions such as language and information exchange, work habits and practices, learning strategies, self-esteem, dress code and professional look, food habits, managing time zone differences, people's values, and beliefs and attitudes related to cultural differences are shown in Figure 1.6.

Communication problems, for instance, in offshoring business from developing countries, may lead to the development of vendor–vendee relationship risk. Stringfellow et al. (2008) in their work defined communication-related invisible hidden costs with offshoring business. D'Mello and Eriksen illustrated information technology outsourcing demands and resulting work culture and attitudes to work. Though outsourcing with virtual teams can help to get the work done, differences in cultural and working habits and work ethics still exist that inhibit teamwork and mutual trust among the stakeholders. In literature, the authors have suggested various means to overcome cultural differences in global

Figure 1.6 Dimensions of risk due to cultural differences.

software development (GSD) domain which includes employing cultural liaisons or mediators (individuals who are familiar with vendor–vendee culture), directing team members to visit other partner's worksite, organizing cross-cultural training sessions, etc.

When a sense of pride is instilled in the offshoring business, members in the team interacting with clients develop more positive emotions, and it becomes easier to work. Self-esteem is not only linked to the knowledge source but also to the knowledge service provider as well. According to Tate and Ellram (2012), for outsourcing the knowledge-oriented services, learning of new skills and methods of performing outsourcing work is a part of the innovation-related task. The above studies are summarized to identify the dimensions resulting from cultural differences. Table 1.4 presents the definitions of these factors.

1.5.3 Opportunistic behaviour risk

The risk arises because of behaviour of the service provider. It may result in lack of coordination followed by shirking, cheating and distorting information. Lim and Tan stated that the vendor presents himself as a competitor

Table 1.4 Cultural difference dimensions with definitions

S. No.	Name of dimension	Definition
1.	Language and information exchange	The nature of communicating language, i.e., verbal or non-verbal differentiates one team from another team and the connotation given to gestures obviously differ from one culture to another culture.
2.	Working habits and work ethics	Working habits and ethics are defined as an effort directed by the workforce to accomplish the tasks.
3.	Learning strategies	It involves how people in teams consolidate and process the project-related knowledge. Some teams favour abstract thinking and some prefer memory and rote learning.
4.	Self-esteem	It is related with the identity of individuals working in diverse cultures. It is demonstrated through bold behaviour in one culture and modest behaviour in another.
5.	Dress code and professional look	This includes dress and costumes worn by teams to present themselves in a culturally distinctive environment.
6.	Food habits	The food habits of teams often differ among cultures as evident from the way in which they prepare, present and eat food of their choice.
7.	Managing time zone differences	In offshoring work, sense of managing time and time difference among persons because of cultural differences. Some cultures exhibit precise time awareness and some exhibit a casual approach.
8.	People's values, beliefs and attitudes	People of all cultures have strong belief and attitude for supernatural power which is apparent from the religious practices followed by them.

once he gains sufficient offshoring knowledge. An opportunistic vendor may offer the services to the firm's competitors, thereby negating the competitive advantage of the firm. The risk of opportunistic behaviour comes into picture when the assets acquired by the organization are used outside the service provider's organization. To address this opportunistic behaviour, a strong commitment at all levels of the organization is required (Søderberg et al. 2013). Adequate training to handle offshore outsourcing projects induces a higher degree of opportunistic risk to client organizations (Kumar et al. 2014).

With the outsourcing of supporting services, viz. information technology services, code development or materials procurement, substantial risk is involved. Trustworthy partners in offshoring business always refrain themselves from opportunistic behaviour. It is observed from the transaction cost theory that trust among partners can lower the risk. In few cases, client organizations opt for reshoring to stay away from the risks of offshoring or to mitigate the adverse experiences of the same in the past. The uncertain outsourcing environment and the information asymmetry in the outsourcing relationship prompts the client organization to act in a way that prevents the service provider organization's opportunistic behaviour. The team members are often exposed to the risk of opportunism and the risk of shirking behaviour. Opportunistic behaviour of the service provider to gain individual benefits at the expense of bad repute results in the failure of the agreement or contract. Many dimensions related to opportunistic behaviour risk (Cai et al. 2011; Gray et al. 2013; Pai 2015; Skowronski et al. 2020) are shown in Figure 1.7.

The widened communication gap between vendor and vendee in the design process is reduced by the participation of a third party in the design

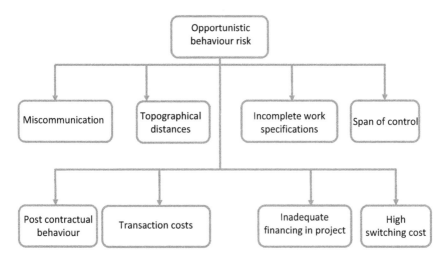

Figure 1.7 Dimensions of opportunistic behaviour risk.

process. Suboptimal output risks result from intricacy of operations, geographical boundaries between vendor and vendee along with communication problems. Offshore outsourcing projects emphasize the existing risks associated with staffing, scheduling and budgeting. This is mostly because offshore outsourcing projects involve cultural and geographical distances. The convenience of establishing an international purchasing office for offshore outsourcing depends on various dimensions like volume of purchase, characteristics of product and geographical distances between the client and service provider organizations.

The issue relates to most clients' apparent lack of knowledge about the skill that is required to perform the task efficiently (Bhattacharya et al. 2013). Incomplete work specifications also pose a greater risk that companies deal with. Moreover, in offshoring, it is difficult to lay down the potential situations in the agreement which may change in the future time period. Client organizations have to embrace certain mechanisms to control risk and address the risk-related dependencies while managing the relationships with suppliers. In offshoring, risk related with loss of critical skills is also faced by the client organization. With increased offshore outsourcing, the breadth and depth of the client organization's dependency grows, often with negative and unanticipated consequences (Ellram et al. 2013).

In offshore outsourcing agreements, lack of appraisal mechanisms, viz. metrics, testing tools and certified quality models, is the most often encountered risk. An issue with offshore outsourcing is the need of real monitoring of the outsourced functions in order to avoid post contractual opportunistic behaviours. Breach of contract and inability to deliver as promised by the service provider poses a risk for outsourcing the services. Partner opportunism has been highlighted in transaction cost economics literature which may affect organizational relationships. The costs of coordinating and monitoring offshored services reduce savings from cheaper labour from low-income countries. Offshoring usually increases the chance of vendors' opportunistic behaviour as evident from under-investment in software development project activities. Handley and Benton Jr (2012) depict the concern that service providers may be inclined to withhold resources or "under-invest" in the relationship if they believe the client organization is unable to detect such actions. Barney et al. (2014) in their work reported that under-investment in projects results in unusable products and over-investment results in more costs than the perceived benefits. Most of the risks are associated with exchanging costs, conversion and management costs, amendment in contract and litigation costs. Due to high switching costs, client organizations become locked into the relationship and this is recognized as an obstacle to offshoring. Client organizations sometimes do not prefer to offshore a highly customized product on a global basis because of high switching costs. Based upon the above studies, dimensions of opportunistic behaviour risk of offshore outsourcing are presented in Table 1.5 with the definitions.

Table 1.5 Opportunistic behaviour risk dimensions

S. No.	Opportunistic behaviour risk dimensions	Definition
1	Miscommunication	The miscommunication or lack of communication between the client and vendor impairs the success offshore outsourcing project and result in delays.
2	Topographical distance	Though topographical distance always exists in offshore outsourcing projects but in IT offshoring services time zone differences between the countries matter and thus timely completion of projects and their delivery performance is affected.
3	Partial work specifications	Incomplete work specifications results when both the organizations are in conflict of goals.
4	Span of control	When span of control with respect to project execution is more lop-sided, which disrupts the functioning of the service provider organization.
5	Pre- & post-contract agreement	Pre- and post-contract behaviour consists of those offshore activities which are to be completed before contract agreement and post-contract agreement includes offshore outsourcing project assessment done after expiration of a contract agreement.
6	Transaction costs	Transaction cost is the cost of specifying what is being exchanged and of enforcing the resulting agreements. These costs vary independently of the competitive market price of goods or services exchanged.
7	Inadequate funding in project	There is a risk of inadequate funding or investment by the service provider to the client organization, not getting what was required as per contract obligations and thus providing the limited benefits.
8	High switching cost	High switching cost is associated with change of service provider organization for offshore outsourcing.

1.5.4 Intellectual property infringement risk

According to the World Trade Organization (WTO), intellectual property rights (IPRs) are "the rights given to persons over the creations of their minds". They are typically given to the creator as an exclusive right over the use of his/her creation for a certain period of time. Under intellectual capital, three different types of capitals, viz., human capital, social capital and organizational capital, which help organizations to attain and sustain competitive advantage, are covered. If intellectual capital is managed properly in an organization, it can lead to superior performance in business. For example, a firm that possesses high-quality engineers/managers in the form of human capital, a trustworthy set of suppliers and/or distributors in the form of social capital and well-documented standard operating procedures in the form of organizational capital is likely to have a superior performance. Firms conduct their business functions either themselves or outsource them to third-party service providers. When selecting the offshore outsourcing service provider, firms often think of trade-offs between loss of IPRs and efficiency. It is observed that firms reap benefits from offshoring even if the likelihood of intellectual property (IP) deceit and moral threat is present.

IP right may be used in many ways, namely, to suppress competition, raise the prices thus increasing the profit, selling the IP right to another enterprise and licensing the IP right. IP risk is a major threat for such innovative product development. Risk of IP is usually highlighted as a major concern when companies deal with suppliers in developing economies (Mihalache and Mihalache 2020). According to Fisher III and Oberholzer-G, only one half of business leaders understand the value of IP. The risk related to IP occurs when the vendors or their staff misuse IP or even where the contract between the parties specifies that IP rights exclusively belong to the clients. In literature, IP risks are termed as copyright risk, patents, trade secrets, trademarks, counterfeiting, piracy, trade dress and industrial labour design right. In the presence of dimensions such as lack of vendor's credentials, managerial problems, and political and economic uncertainties in the offshore country, outsourcing with offshore parties becomes more risky than domestic parties. Also, the costs associated with stolen IP pose a greater challenge. The strategic dimensions (IP, development and trust), along with operational issues (communication, coordination, quality, cost and delivery), make trade-offs between outsourced and domestic development more challenging. If the IP rights of new technology are unprotected, it may result in making inter-organizational innovation a high risk and least profitable. The present concerns related to the IPR protection in software production in India are an early indication of such types of risks. The recent experiences of IT managers have tapped into a new fear, as they began to question the security of IPR property. It has become expensive to train employees dealing with offshore suppliers, which leads to risks associated with supplier turnover and loss of IP. Ambos and Ambos (2011) in their work focused on the negative effects associated with cultural differences among the parties. They concluded that the more the cultural distance between the domestic country and the foreign country, the less likely the firms to establish a knowledge-seeking (versus knowledge-exploiting) partnership.

Leakage of confidential data and loss of IP rights pose a great challenge in offshore outsourcing process. As compared to onshore outsourcing, offshore outsourcing practice should handle more uncertainties and potential risks because of the presence of global challenges owing to dimensions related to culture, linguistics, procedures and laws, security and IPR protection. The protection of IPR rights is a key challenge being faced by companies engaged in outsourcing business. By segmentation of various supply chain stages including production activities along the supply chain and transferring them at different geographic locations, firms may take internalization advantages and prevent losses. Important dimensions related to IP risk are shown in Figure 1.8.

Due to IP protection issues, the details of the projects cannot be furnished and it affects the performances of the projects. There are many intricacies and complexities of distributed software development projects and one of the hurdles is IP risk. There is high demand for software engineers who work in offshoring projects and have strong liaison with customers, skill to manage teams and improve business processes in creative manner, but one of the major issues that needs attention is the institution of IP rights.

IP protection of the host country is related to the presence of institutional norms that can help businesses to protect their rights related to proprietary knowledge. The risk due to IP arises when outsourcing activities necessitate the service providers to gain access to an organization's crucial product technologies in order to commence the outsourcing project. It is important to safeguard the internal knowledge and its output. Indeed, this is the precondition for increasing the level of knowledge, and at the same time if no protection system is in place, then the growth probability would be negative and IPR rights provide necessary safeguards to preserve technical knowledge. During negotiations, organizations usually refrain from mentioning the laws related to IP in their country. The organizations thus ensure that service providers shall not operate in a lawless environment.

A client organization might suffer if its IP is badly handled by its service providers. Opportunistic behaviour by a service provider emerges when a client organization reveals valuable knowledge and information to a service provider, or facilitates access to it. The use of the internet provides

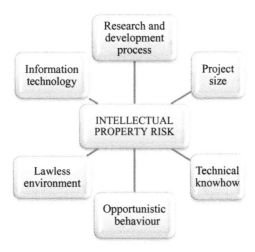

Figure 1.8 Dimensions of intellectual property infringement risk.

communication and commerce business to take place with or without borders, given significant problems in investigation and enforcement of economic offences. Researchers suggested that the knowledge transfer in offshoring business across the countries may expose firms to various risks. The critical challenge being faced by IP decisions is the substantial risk of leaking confidential information through multiple service providers and clients (Pisani and Ricart 2018). Based upon the above review of literature, dimensions of IPR risk of offshore outsourcing are presented in Table 1.6 with definitions.

1.5.5 Financial risk

The cost associated with managing the complexity of risks involved in outsourcing should be included in the outsourcing process. Offshore outsourcing reduces some costs, but at the same time it results in certain expenses which are related to supplier selection costs, legal/contract costs and the transition costs incurred on outsourcing work to other parties. The changeover period is the costly stage in offshore outsourcing. Some vendors try to elicit more profit from existing contracts or provide additional services to increase returns. According to Chou and Chou, and Ray et al. (2013), various cost-associated risks include unpredicted switching costs, costly contractual amendments and litigation costs. When the supply chain of a company is disrupted, the global networking of operations

Table 1.6 Intellectual property infringement risk dimensions with definition

S. No.	Intellectual property infringement risk dimensions	Definition
1	Project size	Project size refers to an innovative task assigned to a global team. Project size estimation is a crucial aspect, as it helps in planning and allocating the resources for a project.
2	Research and development process	In offshore outsourcing, service provider develops research and development process for client organization.
3	Technical knowhow	For performing the specified task, client organization gives knowledgeable inputs to the service provider.
4	Lawless environment	Lawless environment exists in the service provider's country if laws related to intellectual property infringement are not implemented properly.
5	Opportunistic behaviour	It refers to the behaviour of service provider if the information related to client organization is used in an unethical manner.
6	Information technology	Use of Information technology as a threat to intellectual property rights of client organization. Intellectual property law is significantly impacted by digital technology, particularly in the fields of copyright, trademark and patent law.

leads to "considerable financial loss" to the company. Dimensions related to financial risk are shown in Figure 1.9.

Ray et al. (2013) presented a decision analysis qualitative approach to financial risk management in strategic outsourcing. Small and medium enterprises with their limited financial power tend to go for offshoring to seek business solutions. Large firms on other hand with sound financial position possess more offshoring advantages in terms of economies of scale and scope. Variables related to closeness with core business, switching and adaptation costs are evaluated for the strategic importance of an offshore outsourced activity by Gunasekaran et al. (2015) and Bruccoleri et al. (2019) in their works. They concluded that the reasons to go for contract renewal are primarily related to the high switching costs. Layoffs or continued voluntary job loss by employees also has a detrimental effect on a company's ability to perform proficiently because most of the time they have to deal with workforce issues. The inability to measure their business performance due to incomplete or non-measurable specifications affects the financial health of the firms (Bruccoleri et al. 2019). The dimensions of financial risk are presented in Table 1.7 with definitions.

1.5.6 Organization structural risk

To coordinate various operations firms establish their organizational structures, the lack of coordination among various entities in an organization results in organizational structural risk (Albrecht 2018). Hence,

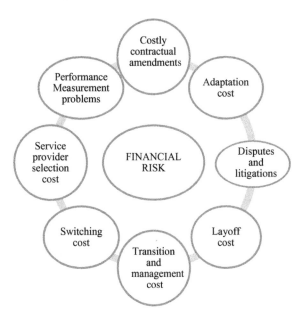

Figure 1.9 Dimensions of financial risk.

Table 1.7 Financial risk dimensions with definitions

S. No.	Dimensions	Definition
1	Switching cost	Alternative cost arising due to change of assigned tasks from one service provider to another.
2	Transition and management cost	These are the costs associated with initializing and start-up activities when tasks are transferred from client organization to service providers.
3	Layoff cost	For economic reasons to cut down running cost of business the employer initiates layoffs in which the company downsizes the workforce.
4	Service provider selection cost	These are the costs which are incurred in selecting a service provider.
5	Performance measurement problems	The conflict of identification/measurement of performance between client organization and service provider.
6	Costly contractual amendments	Such cost arises out of revision/amendments which are not identified in the initial contract between the client and the vendor.
7	Disputes and litigations	Being geographically and culturally apart creates problems which may result in disputes and litigations resulting in such costs.
8	Adaptation cost	Adaptation cost is associated with the competency which is being developed by the client organization for understanding the high-level solutions which are provided by the service provider.

coordination among various entities in an organization involved in business is imperative to achieve effectiveness in operations related to various offshoring services. In an organization, specialization, standardization, and vertical and horizontal integration are significant dimensions in determining the outsourcing decisions. Most firms do not worry about the behaviour of clients when they enter into the contract with them. It is assumed by them that vendors act in ways that maximize interests of both groups. However, third-party service providers make alterations with respect to process, technology and operating procedure without taking the clients into confidence. The key asset management issues include investments made by vendors on training, modifying processes to accommodate clients' systems, investments being made, etc. As outsourcing often entails a long-term contract which seldom takes into account future eventualities, it becomes crucial for the service providers to adapt to capacity, both in geographical and technical manners, in a changing business environment. The activities might call for implicit knowledge which is not easy to measure, and as a result it complicates the offshoring issues among the parties. The mid-contract problem often occurs when the vendor dispenses all transformational aspects, viz., consolidation, standardization, superior technology and improved process.

Client organizations observe another kind of risk when vendors alter the agreement terms after clients outsource their business processes to them for completion, and it occurs because, as work according to outsourcing agreements is near completion, the change in relationship shifts from buyers to sellers (Ray et al. 2013). Many dimensions related to organization structural risk are shown in Figure 1.10.

Organizations must strike a balance between potential returns and country-specific risks that depends on change in the political environment, and regulatory and economic conditions. Rao (2004) in their work discussed issues about performing business overseas and identified dimensions, viz. telecommunication infrastructure availability, difference in language and legal issues in conducting offshoring assignments. Outsourcing may also result in dropping employee morale. Winkler et al. (2008) suggested that conflict among the parties because of the mismatch of activities and goals has a negative impact on the business. The more explicit the knowledge, the more challenging it is to find and train the workforce and the longer the delays in accomplishing the tasks on time. On the other hand, projects with flexible schedule agreements pose risk related to additional costs and unexpected delays. Once a project is in place, it becomes difficult for the vendor organization to train and maintain the staff. Vendor technical expertise, contract agreement and project handling; physical location among the parties; the number of customers; and transaction costs are key challenges in offshore outsourcing. Introduction of Information technology in an offshore environment

Figure 1.10 Dimensions of organization structural risk.

provides success in service delivery. According to Cappelli (2011), inability of firms to preserve human capital also acts as major critical risks in offshore outsourcing. The tools that protect an organization include policy documents and operating procedures that organizations adopt to prevent threats and manage reclamation. Table 1.8 presents the details of dimensions along with definitions related to organization structural risk.

1.5.7 Operational risk

Operational risk arises when services are not delivered as per contract or performance issues are observed because of operational failures related to infrastructure or technology. This type of risk results in suboptimal output which is obtained as a result of complex operations, geographical distance between vendor and client, and the limitations because of language differences between parties (Dolgui and Proth, 2013; Chauhan et al. 2015; Rahman et al. 2020). Some smart firms kick-start their offshore activities with an assumption that service providers may not be in a position to execute their processes for a long time. The transition phase arrangement had shown poor results in terms of project completion both for client as well as service providers. In addition, measurement problems arise which are related to difference in interpretation of service performed which usually happens when things are subjective. When the services offered by a third party are not standardized, both the quality of service

Table 1.8 Dimensions of organization structural risk with definition

S. No.	Risk dimensions	Definition
1	Regulatory issues	Regulatory issue arises when an outsourcing agreement obstructs the client from necessary compliance with the regulatory system in place.
2	Incompatibility	Client and service provider organization are very different by nature and mismatch between their working styles results in incompatibility issues.
3	Delay in completion of project	Delay can be defined as a later execution than the intended time limit, or particular period, or later than the specific time period that all the concerned organizations had agreed to for execution of the project.
4	Inappropriate resource allocation	Most service providers do not take into consideration all the available possibilities and consider only organizational resources that are not appropriate to accomplish their goals.
5	Deficiency in capabilities	Deficiency in capabilities is the risk which arises when the service provider doesn't have the international dimension or past experience that is needed in order to evolve together with the client organization.
6	Reduction in human capital	Reduction of human capital possessing the right skill set and competencies to accomplish the assigned task.

and its reliability result in substantial risk and it normally happens when knowledge process work is outsourced. The alleviation of risk associated with operations in outsourcing is very serious because the cost of failure of contract is high due to costs involved in the project. Numerous times this risk comes to the fore when the client loses control over operations due to offshore outsourcing. Monitoring the offshore performance can be challenging owing to the difficulties faced during coordinating and communicating with offshore vendors. Agencies providing outsourcing services often have to fix erroneous enterprise processes later on, if not identified in the initial stage. Such operational risks in a later stage can have an undesirable effect on vendors' performance and reduces their profit margins with vendors' inability to deliver per promise being a critical issue in outsourcing. Many dimensions related to operational risk are shown in Figure 1.11.

The main dimensions that affect the offshore outsourcing process are service providers' technical competency, reliability and financial strength. Because of greater fragmentation in business processes across the supply chains, a standard operating procedure which includes use of standard technology or standard processes does not develop. Thus, it has become subsequently more challenging for organizations to cope up with heterogeneous data, business processes and technologies by service providers

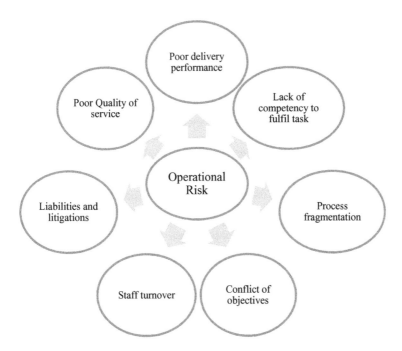

Figure 1.11 Dimensions of operational risk.

and competitors. It is evident that various traditional monitoring and control procedures are not feasible in offshore outsourcing assignments, and the greater amount of virtual collaboration between the parties may lessen the chances to solve conflicts among clients and service providers and make it more challenging to build long-term relationships (Søderberg et al. 2013). Contracts between outsourcing service providers and clients are always delicate, potentially adversarial and sometimes the scene of outright conflict. Also, offshore outsourcing is related to high employee turnover and inadequate managerial experience which results in late delivery of outsourcing services. The limitation with the collaboration model is financial risk, i.e., adoption of the vendor–vendee collaboration model may raise concerns about the malpractices because of negligent representation. Outsourcing normally reduces a firm's control over the way the services are delivered, which in turn may raise the firm's liability confessions. Offshore contracts can result in distributed environment development, which is also linked to work-related quality issues If time zone difference is greater, then the adverse effect on client quality of service is also greater. Bad quality of service and high cost with offshore contracts may result in early contract terminations, and switching by companies to IT insourcing or developing their own IT capabilities (Andrea et al. 2016). Dimensions of operational risk are presented in Table 1.9 with definitions.

Table 1.9 Operational risk dimensions with definitions

S. No.	Risk	Definition
1	Poor delivery performance	Poor delivery performance is when services are not delivered as expected or not in time.
2	Lack of competency in fulfil task	Lack of competency can be considered as not having required characteristics for performing a given task, activity or role successfully.
3	Process fragmentation	Many problems arise due to complex process of offshore outsourcing that spans along various dimensions between the client and service provider.
4	Conflict of objectives	Differences in the interest of both service provider and clients as they work as an individual firm on the basis of local perception and opportunistic behaviour may result in conflict of objectives.
5	Staff turnover	It is the proportion of employees who leave an organization during a certain period of time.
6	Liabilities and litigations	They arise from lack of clear lines of responsibility and accountability for the service provider organization.
7	Poor quality of service	Quality of service of a task is deteriorated due to poor engineering support, lack of sharing of new technology and lack of service promptness by the service provider.

1.5.8 Cybersecurity risk

The UK Office of Science and Technology defines cybersecurity as "defences against electronic attacks launched via computer systems". Terms such as "IT security event", "cybercrime" and "cyber-event" refer to the concept of risk in the cyber environment. In a traditional supply chain, there is flow of goods, finance and information, whereas in a cyber supply chain, IT infrastructure and technologies are extensively used to connect, build and share data among entities in a supply chain network. Cyber supply chains add complexity to business and have become more challenging to manage. Today, firms have become aware of risks associated with cybersecurity and hence they have raised their budgets to counter the risks (KPMG 2017). The risk survey conducted globally by Gartner, AXA, Society of Actuaries, Deloitte firms in 2018 revealed that cybersecurity and data outages appeared as the one of the top risks which modern enterprises face today. According to Xue et al. (2013), Urciuoli (2015), and Urciuoli and Hintsa (2016), the studies in literature do not address the consequences of cyberthreats satisfactorily in the context of supply chain at different levels.

> Information security poses a huge challenge in business as much of our information services are outsourced. However, some important challenges also need to be considered, such as lack of data, insiders, IT vulnerabilities, regulatory frameworks, criminal behaviour, etc. Hence, recommendations are made for managers to improve their understanding of supply chain security.
>
> *(Urciuoli and Hintsa 2016)*

According to a World Bank report, 40% of business firms consider security as a primary obstacle in adopting cloud services because of issues related to data privacy, compliance and access controls, lack of trust and transparency and shared responsibility. These concerns on cybersecurity are more for smaller business firms as they outsource more of their IT needs. According to a Global report by Ponemon Institute (2021), the cybersecurity incidents caused by insiders have increased by 47% since 2018. For instance, a former engineer of amazon web services was found guilty of stealing the personal information of 100 million customers linked to Capital One in 2019. Other examples of IT outsourcing risk are Salesforce's multi-hour cloud meltdown because of error in database which grants the users right to use (May, 2019); in June, 2019, Google's cloud outage brought down services of YouTube, Gmail and Snapchat in various regions of the United States.

The growing capability of cloud-based IT systems have advanced the threats related to cybersecurity, thus making cybersecurity risk concerns dominant in all forms of IT outsourcing business. While outsourcing their

IT needs, firms unequivocally presume that IT outsourcing service providers abide by their concerns for risks associated with cybersecurity. In literature, cybersecurity-related risks are an emerging topic in the context of supply chain risk management, which calls for attention from supply chain and cybersecurity managers. Various dimensions related to cybersecurity risk are shown in Figure 1.12. The main dimensions summarized from the studies undertaken by Bomhard and Daum (2021), Ghadge et al. (2019), Urciuoli (2015), Peck (2006) are related to cross-country risk, within-country risk, physical threats, Software vulnerabilities, cyber assaults or attacks, insider threats because of human resources and contextual risk factors. Table 1.10 presents definitions of various cybersecurity risks.

1.6 CONCLUSION

In the last two decades, various empirical as well as conceptual methods are available in literature where researchers have studied or modelled the effect of various risks in offshore outsourcing environment. Today business offshore outsourcing has become a way of life and organizations are increasingly using outsourcing to streamline workflow and save costs (BY SITEL STAFF OCTOBER 21, 2021). This chapter presents details of offshore outsourcing and its definitions. The risks related to offshore outsourcing and risk drivers are defined along with their key drivers with inputs from literature.

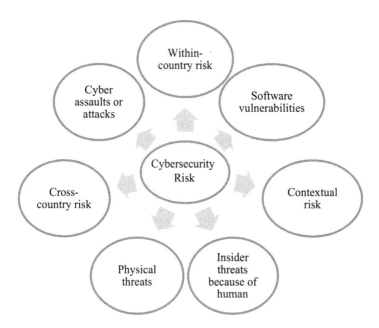

Figure 1.12 Dimensions of cybersecurity risk.

Table 1.10 Cybersecurity risk dimensions

Dimensions	Sub-dimensions	Description
Physical threats	ICT devices Control mechanisms	These types of risks affect the supply chain functions and security of physical infrastructure components. The physical infrastructure includes switches, servers, routers, firewalls and other ICT devices. For instance, when a natural disaster occurs, then the information flow among the entities in the supply chain network is disrupted.
Software vulnerabilities	Outdated subsystems	To execute different tasks, entities in the supply chain depend on different software systems. These vulnerabilities are embedded into the software during the design or implementation phase. The vulnerabilities are discovered or updated on a continual basis before making public for intended users. On the other hand, some of the systems are neither updated nor fixed at all. In both cases, they are extremely vulnerable.
Cyber assaults or attacks	Direct attacks Indirect attacks	The risk due to cyberattacks is direct and indirect. Virus and hacking attacks comes under direct attacks. They impact operations which consist of risk associated with industry spying or infringement of IPR. Under indirect type of attacks, the hackers lay out a "bait" to breach the security of systems.
Insider threats because of human resources	Deliberate Premediated	Insider threats are caused by workers in which they use their authorized access to harm the organization. For example, insiders could provide substantial support to external hackers by revealing weak points, or disclosing authenticated passwords. Whether the cyber breach by an employee is deliberate or accidental or negligent, it is termed an insider threat. Thus, the human factor can pose the biggest threat to a firm's cybersecurity.
Within-country risk	Government regulations Legal standards	Within-country risk sources involve regulatory rights or government policies on digital connectivity, digital e-commerce and digital intellectual property. It also entails riders against foreign companies towards digital connectivity. Legal standards that prevent internet frauds fall under within-country risk category.
Cross-country risk	Geopolitics Bilateral relations	Cross-country or geopolitics risk exacerbates when bilateral relations between home and host country worsens. Because of divergent ICT standards between economies (United States vs China norms) related to digitization systems, the cross-country risks pose implications for national security. Moreover, the digital supply chains make firms more prone to cyber risks.
Contextual risks	Procurement Staffing Funding Training IT infrastructure	The contextual risks are associated with supporting processes such as procurement, staffing, funding, training and development, and the digital infrastructure viz. automated processes, strong AI-supported algorithms, and cloud computing services.

Because of the multifaceted nature of offshoring services and its linkage with cultures, disciplines and technologies it has become important to analyse the association among the various risks. Such analysis will help answer the following questions:

- How cultural changes impact the management of geographically distributed overseas business operations?
- How to uphold and improve the quality of offshore operations?
- How to reduce opportunistic behaviour of service providers in offshore outsourcing?
- How to tackle geo-political environment for offshore operations which help to mitigate risk?
- How to safeguard the supply chains with cybersecurity risks?

The upcoming chapters in the book will analyse the various risks such as culture differences, opportunistic behaviour risk, political risk, IP infringement risk, financial risk, compliance and regulatory risk, organization structural risk, cybersecurity risk, etc., which have considerable impact on offshore outsourcing.

SUGGESTED READINGS

Chauhan, P., Kumar, S., Sharma, R.K. (2017). Investigating the influence of opportunistic behaviour risk factors on offshore outsourcing. *International Journal of Business Excellence*, 12(2), 249. https://doi.org/10.1504/ijbex.2017.083570.

Dolgui, A., Proth, J.-M. (2013). Outsourcing: definitions and analysis. *International Journal of Production Research*, 51(23–24), 6769–6777. https://doi.org/10.1080/00207543.2013.855338.

Gupta, R.S. (2018). Risk management and intellectual property protection in outsourcing. *Global Business Review*, 19(2), 393–406. https://doi.org/10.1177/0972150917713536.

Ishizaka, A., Bhattacharya, A., Gunasekaran, A., Dekkers, R., Pereira, V. (2019). Outsourcing and offshoring decision making. *International Journal of Production Research*, 57(13), 4187–4193. https://doi.org/10.1080/00207543.2019.160369.

Kaur, H., Singh, S.P., Majumdar, A. (2019). Modelling joint outsourcing and offshoring decisions. *International Journal of Production Research*, 57, 4278–4309. https://doi.org/10.1080/00207543.2018.1471245.

König, A., Spinler, S. (2016). The effect of logistics outsourcing on the supply chain vulnerability of shippers. *The International Journal of Logistics Management*, 27(1), 122–141. https://doi.org/10.1108/ijlm-03-2014-0043.

Mihalache, M., Mihalache, O.R. (2020). What is offshoring management capability and how do organizations develop it? A study of Dutch IT service providers. *Management International Review*, 60, 37–67. https://doi.org/10.1007/s11575-019-00407-5.

Nordås, H.K. (2020). Make or buy: offshoring of services functions in manufacturing. *Review of Industrial Organization, 57*, 351–378. https://doi.org/10.1007/s11151-020-09771-1.

Pisani, N., Ricart, J.E. (2016). Offshoring of services: a review of the literature and organizing framework. *Management International Review, 56*(3), 385–424.

Rao, M.T. (2004). Key issues for global IT sourcing: country and individual factors. *Information Systems Management, 21*(3), 16–21.

Uygun, Y., Nikoloz, G., Florian, S., Lizi, G., Brigitte Stephanie, T.N. (2023). A holistic model for understanding the dynamics of outsourcing. *International Journal of Production Research, 61*(4), 1202–1232.

Chapter 2

Qualitative structural models for various offshore outsourcing risks

2.1 INTRODUCTION

When outsourcing activity is accomplished through offshore vendors, it results in various risks in terms of language barriers, time zones, cultural differences and geographical barriers. This chapter presents the qualitative structural models for various types of risks discussed in Chapter 1, i.e., political risk, risk due to culture differences, opportunistic behaviour risk, intellectual property infringement risk, financial risk, organization structural risk and operational risk. The qualitative models are developed for each type of risk to understand the structural relationship among key variables in various offshore outsourcing risks and further, the MICMAC analysis has been used to classify the risks and validate the interpretive structural model. MICMAC categorizes the risks under four main categories: (i) independent risks, (ii) linkage risks, (iii) dependent risks and (iv) autonomous risks.

The different steps, which show the way to develop the qualitative model using the interpretive structural modeling (ISM) approach, are discussed as under.

Step 1: Determining the dimensions of interest relevant to the problem or issues by critical scrutiny of literature or using the group problem-solving technique.

Step 2: Developing a contextual relationship between the identified dimensions according to which various pairs of dimensions need to be examined.

Step 3: Development of SSI (Structural Self-Interaction) matrix to indicate pair-wise relationship among the dimensions.

Step 4: Development of a reachability matrix (RM) from SSI matrix, and examining it for transitivity. Transitivity refers to the assumption that, if dimension "*A*" is influencing "*B*", and "*B*" is influencing "*C*", then "*A*" will be necessarily influencing "*C*".

DOI: 10.1201/9781032707884-2

Note: The SSI matrix is changed into a binary matrix, called the initial RM by replacing *V*, *A*, *X* and *O* with 1 and 0 based on the following arrangement:

- If (i, j) value in SSI matrix is represented by *V*, then in the RM, (i, j) is represented by 1 and (j, i) is represented by 0.
- If (i, j) value in the SSI matrix is represented by *A*, then in the RM, (i, j) is represented by 0 and (j, i) is represented by 1.
- If (i, j) value in the SSI matrix is represented by *X*, then in the RM, (i, j) and (j, i) both are represented by 1.
- If (i, j) value in the SSI matrix is represented by *O*, then in the RM (i, j) and (j, i) both are represented by 0.

Step 5: The RM is split into different levels.

Step 6: A directed graph called diagraph is drawn based on the relations in RM, and links with transitive relations are removed.

Step 7: In this step, the digraph is changed to a hierarchy model ISM by replacement of nodes with dimension statement.

Step 8: In the last step, the ISM model is reviewed and is checked for conceptual inconsistences, if any.

Figure 2.1 presents the step by step overview of the ISM approach being used for developing qualitative models for various offshore outsourcing risks.

The basic terms used to denote the association between dimensions is shown by various notations. The four notations, viz. *V*, *A*, *X* and *O* with their relationship meanings as shown in Table 2.1 are used to showcase the association among the dimensions (i and j).

2.1.1 MICMAC analysis

Matrice d'Impacts Croisés Multiplication Applied to a Classification is an expanded form for MICMAC. The MICMAC analysis is done on the findings of the ISM model. The MICMAC is performed to study the driving and depending power of the dimensions of interest. The dimensions are grouped into four quadrants as per their powers of driving and dependence, discussed as under:

1. *Autonomous dimensions:* Those dimensions possessing the weakest powers of driving and dependence fall in the group of autonomous dimensions.
2. *Linkage dimensions:* The dimensions having the strongest powers of driving and dependence falls in the group of linkage dimensions.
3. *Dependent dimensions:* Dependent category consists of the dimensions which have weak power of driving and high power of dependence.
4. *Independent dimensions:* Here, in this category, dimensions possess strong power of driving and weak power of dependence.

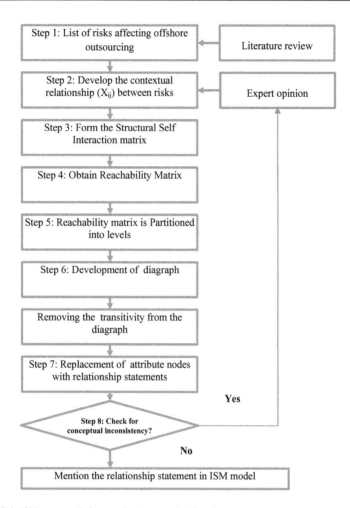

Figure 2.1 ISM approach for qualitative model development.

Table 2.1 Direction of relationship between dimensions

Notation	Meaning
V	Dimension *i* influences dimension *j*.
A	Dimension *j* influences dimension *i*.
X	Both dimensions *i* and *j* are influenced by each other.
O	Dimensions *i* and *j* will not influence each other.

2.2 POLITICAL RISK AND ITS DETERMINANTS

As discussed in Section 1.5.1, offshore destination political uncertainty is one of the important risks which affect the parties involved in offshore outsourcing business. The risks associated with political uncertainty are not only volatile but also more difficult to observe. Also, as per summarized views of respondents [Annexure Table A2] from selected business process outsourcing organizations, the political risk may be due to the number of determinants as (i) civil disturbance, (ii) fiscal and monetary policy, (iii) terrorist events, (iv) industrial labour relations, (v) public policy of the host country, (vi) corrupt behaviour and bribery, (vii) foreign affairs and (viii) availability of suitable human resource.

In order to understand the association among these variables, it is necessary to model these factors for which interpretive structural model has been developed. The details of the model so developed are discussed in the upcoming paragraphs.

2.2.1 Qualitative structural model

Following the steps presented in Figure 2.1 and discussed in Section 2.1, a qualitative structural model has been developed. Based on the contextual association among various dimensions, the SSI matrix is developed. To arrive at a consensus, the SSI matrix so developed is discussed with experts from industry and academia. On the basis of the answers from respondents, the SSI matrix was constructed and is presented in Table 2.2.

Table 2.3 presents the details of RM showing driving and dependency powers against each dimension of political risk.

After creation of the final reachability matrix (FRM), further processing is done to obtain the hierarchical model based on association among the dimensions. To do this, the reachability set and antecedent set for each

Table 2.2 SSI matrix for political risk dimensions

S. No.	Dimensions	2	3	4	5	6	7	8
I	Civil disturbance	O	X	V	A	O	V	A
2	Fiscal and monetary policy		O	O	A	O	V	O
3	Terrorist events			O	A	O	V	V
4	Industrial labour relations				A	O	V	A
5	Public policy of host country					V	V	O
6	Corrupt behaviour and bribery						V	O
7	Foreign affairs							O
8	Availability of suitable human resource							

dimension are obtained for the identified dimensions and associated levels. The dimensions for which both sets of reachability and sets of intersection are the same are placed at the topmost level in the ISM model. In this case, the level identification is completed using four iterations for the eight dimensions of political instability. All the four iterations are shown in Table 2.4.

Based on the final result of iterations, an initial diagraph is obtained which includes transitivity links. After eliminating the transitivity links, a diagraph is drawn as shown by Figure 2.2.

Finally, as per the procedural steps shown in Figure 2.1, the diagraph showing risk dimensions is transformed into an interpretive structural model by replacement of nodes with nomenclature of political risks as presented in Figure 2.3.

Table 2.3 Final reachability matrix of political risk factors

S. No.	Dimensions	1	2	3	4	5	6	7	8	Driving power
1	Civil disturbance	1	0	1	1	0	0	1	1	5
2	Fiscal and monetary policy	0	1	0	0	0	0	1	0	2
3	Terrorist events	1	0	1	1	0	0	1	1	5
4	Industrial labour relations	0	0	0	1	0	0	1	0	2
5	Public policy of host country	1	1	1	1	1	1	1	1	8
6	Corrupt behaviour and bribery	0	0	0	0	0	1	1	0	2
7	Foreign affairs	0	0	0	0	0	0	1	0	1
8	Availability of suitable human resource	1	0	1	1	0	0	1	1	5
	Dependence	4	2	4	5	1	2	8	4	

Table 2.4 Iterations for political risk dimensions

Dimensions	Reachability set	Antecedent set	Intersection set	Level
Iteration I				
1	1,3,4,7,8	1,3,5,8	1,3,8	
2	2,7	2,5	2	
3	1-3-4-7-8	1-3-5-8	1-3-8	
4	4,7	1,3,4,8	4	
5	1,2,3,4,5,6,7,8	5	5	
6	6,7	5,6	6	
7	7	1,2,3,4,5,6,7,8	7	1
8	1,3,4,7,8	1,3,5,8	1,3,8	

(Continued)

Table 2.4 (Continued) Iterations for political risk dimensions

Dimensions	Reachability set	Antecedent set	Intersection set	Level
Iteration II				
1	1,3,4,8	1,3,5,8	1,3,8	
2	2	2,5	2	II
3	1,3,4,8	1,3,5,8	1,3,8	
4	4	1,3,4,8	4	II
5	1,2,3,4,5,6,8	5	5	
6	6	5,6	6	II
8	1,3,4,8	1,3,5,8	1,3,8	
Iteration III				
1	1,3,8	1,3,5,8	1,3,8	III
3	1,3,8	1,3,5,8	1,3,8	III
5	1,3,5,8	5	5	
8	1,3,8	1,3,5,8	1,3,8	III
Iteration IV				
5	5	5	5	IV

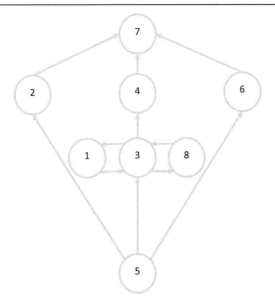

Figure 2.2 Diagraph of dimensions of political risk.

2.2.2 Model analysis and validation

To assess the power of dimensions, viz., driver and dependent power and to further classify them under different clusters, namely, autonomous, linkage, dependent and independent, MICMAC analysis as discussed in Section 2.1

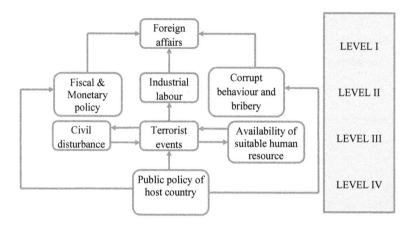

Figure 2.3 Political risk dimensions in the ISM model.

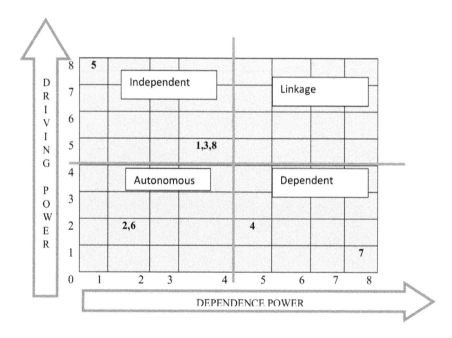

Figure 2.4 MICMAC diagram for political risk.

is used. Figure 2.4 presents the results of the MICMAC analysis which is used to validate the ISM for political risk.

As per the MICMAC analysis, it is found that dimensions, viz. public policy of host country, civil disturbance, terrorist events and availability of suitable workforce are strong drivers and possess less dependence on other dimensions. Uncertainty in rules and regulations is generally considered as a risk because it increases many hidden costs related to the project.

Domestic conflict reduces the intellectual stability of the employee of the host country and it deteriorates their decision-making capacity. People are not easily available for the inspection and other monitoring related tasks in such countries which face the problem of terrorist events. Issues related to employees are treated differently in different countries and many times it reduces the speed of the work.

From the MICMAC diagram (Figure 2.4), one may observe that industrial labour relations and foreign affairs possess the weakest driving power but strong dependence power. Intermittent type of outsourcing work for the service provider organization's employee creates dissatisfaction among them and it deteriorates output of the work. Cultural difference between the nations also reflects the gap between the relationship and it creates distances between the organizations.

Fiscal and monetary policy and corrupt behaviour and bribery fall under the autonomous category, and also they are observed as both weak drivers and weak dependents. Imposition of sanctions creates chaos in the collaborative work. Many times client organizations do not know about the nature of the practices run in the country of the service provider organization and later on, due to this, the overall cost is increased.

2.3 RISK DUE TO CULTURAL DIFFERENCES AND ITS DETERMINANTS

As discussed in Section 1.5.2, difference in cultures among the parties involved in offshoring is often related to attitudes and beliefs which are even more difficult to witness than differences due to language, so they may go unobserved. Also, as per summarized views of respondents from selected business process outsourcing organizations [Annexure Table A3], the cultural differences risk may be due to a number of determinants such as (i) language and information exchange, (ii) working habits and work ethics, (iii) learning strategies, (iv) self-esteem, (v) dress code and professional look, (vi) food habits, (vii) managing time zone differences and (viii) people's values, beliefs and attitudes.

In order to understand the relationship among these variables, it is necessary to model these factors for which the interpretive structural model has been developed. The details of the model so developed are discussed in the upcoming paragraphs.

2.3.1 Hierarchal structural model

Following the steps presented in Figure 2.1, the hierarchal structural model has been developed. Based on the contextual association among various dimensions, the SSI matrix is developed. To arrive at a consensus, the SSI matrix so developed is discussed with experts from the industry and academia. Based on the responses, the SSI matrix was finalized and is shown in Table 2.5.

Following the procedure for ISM model development, FRM is constructed and powers of driving and dependence are shown in Table 2.6 and it takes into account the transitivity concept as per the of ISM procedure.

After creation of the FRM, further processing is done to obtain the hierarchical model based on association among the dimensions. To do this, the sets of reachability and antecedent for each dimension are obtained for the identified dimensions and associated levels. The dimensions for which the sets of reachability and the intersection are similar are placed at the top of the ISM model.

In this case, the level identification is completed using three iterations for the eight dimensions of risk due to cultural differences. All the three iterations with corresponding levels are shown in Table 2.7.

Based on the final result of iterations, an initial diagraph is obtained which includes transitivity links. After removal of the transitivity links, a diagraph is obtained as shown in Figure 2.5.

Table 2.5 SSIM of factors of risk due to cultural differences

S. No.	Dimensions	2	3	4	5	6	7	8
I	Language and information exchange	O	V	O	O	O	O	A
2	Working habits and work ethics		X	A	O	O	X	A
3	Learning strategies			X	O	O	V	A
4	Self-esteem				O	O	O	A
5	Dress code and professional look					O	O	A
6	Food habits						O	A
7	Managing time zone differences							A
8	People's values, beliefs and attitudes							

Table 2.6 Final reachability matrix of risk due to cultural differences

S. No.	Dimensions	I	2	3	4	5	6	7	8	Driving power
I	Language and information exchange	I	I	I	I	0	0	I	0	5
2	Working habits and work ethics	0	I	I	I	0	0	I	0	4
3	Learning strategies	0	I	I	I	0	0	I	0	4
4	Self-esteem	0	I	I	I	0	0	I	0	4
5	Dress code and professional look	0	0	0	0	I	0	0	0	I
6	Food habits	0	0	0	0	0	I	0	0	I
7	Managing time zone differences	0	I	I	I	0	0	I	0	4
8	People's values, beliefs and attitudes	I	I	I	I	I	I	I	I	8
	Dependence	2	6	6	6	2	2	6	I	

Table 2.7 Iterations of factors of risk due to cultural differences

Dimensions	Reachability set	Antecedent set	Intersection set	Level
Iteration I				
I	1,2,3,4,7	1,8	I	
2	2,3,4,7	1,2,3,4,7,8	2,3,4,7	I
3	2,3,4,7	1,2,3,4,7,8	2,3,4,7	I
4	2,3,4,7	1,2,3,4,7,8	2,3,4,7	I
5	5	5,8	5	
6	6	6,8	6	
7	2,3,4,7	1,2,3,4,7,8	2,3,4,7	I
8	1,2,3,4,5,6,7,8	8	8	
Iteration II				
I	I,	1,8	I	II
5	5	5,8	5	II
6	6	6,8	6	II
8	1,5,6,8	8	8	
Iteration III				
8	8	8	8	III

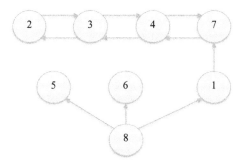

Figure 2.5 Diagraph of dimensions of risk due to cultural differences.

From the diagraph, to obtain the ISM model, nodes are replaced with name of the risk dimensions related to differences in culture as presented in Figure 2.6.

2.3.2 Model analysis and validation

To classify the dimensions and check their driving and dependence power, the MICMAC analysis has been done. Also, it is used to validate the hierarchal model showing the dimensions of cultural differences in the study. The quadrants (i)–(iv) in Figure 2.7 shows dimensions related to cultural differences obtained from the MICMAC analysis.

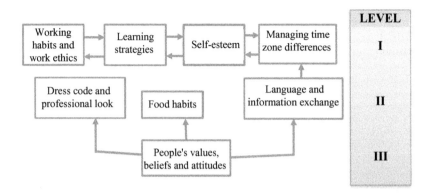

Figure 2.6 ISM model of dimensions of risk due to cultural differences.

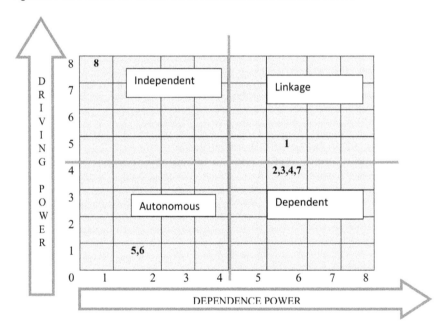

Figure 2.7 MICMAC diagram of risk dimensions due to cultural differences.

The MICMAC diagram presented in Figure 2.7 shows that people's values, beliefs and attitudes dimension possess the strongest driving power and less dependence power. It may happen that few people believe that they are superior to others and it creates problems within the group about due respect to each other and it creates chaos. Many times, it happens that people from one culture have less knowledge about people's values, beliefs and attitude of other cultures, and it creates an unpleasant scenario for operation managers.

The cultural differences risk is the linkage dimension with strong power of driving as well as and dependence. Due to the pronunciation problem, misunderstanding may be created among the employees of different organizations. On many occasions, in a discussion, few resources speak in their native language and it has a negative impact on the others in the group.

It is perceived that dimensions, viz. working habits and work ethics, managing time zone differences, self-esteem and learning strategies possess weak driving powers but at the same time these dimensions exhibit high dependency. Sometimes, it happens that people of one culture enjoy the work in autonomy or individually, whereas others like working in a group and there is a mismatch in their preferences. Intellectual learning strategies among teams usually occur generally in an informal environment. Many times, people feel uneasy in the environment of a different culture. There is a mismatch between the culture of structured and formal type versus culture of more flexible and informal type, and it creates problems among all the resources of different organizations. Day-to-day activity cannot be monitored from a distance, and many times, it reflects on the results. The MICMAC diagram with four quadrants (Figure 2.7) shows that the dimensions dress code and professional look along with food habits are observed in an autonomous category. Most of the times, even colour of dress is wrongly interpreted by other persons and it creates unpleasant situation. Eating habits also may differ between cultures and due to this, people feel uneasy and their efficiency is also not up to the mark.

2.4 OPPORTUNISTIC BEHAVIOUR RISK AND ITS DETERMINANTS

As discussed in Section 1.5.3, opportunistic behaviour risk is associated with the behaviour of offshoring service providers. It may effect coordination among the parties involved and also comprises eluding, cheating and misrepresenting information also as per summarized views of respondents from selected business process outsourcing organizations [Annexure Table A4]; this risk may be due to a number of determinants as (i) miscommunication gap, (ii) topographical distances, (iii) partial work specifications, (iv) span of control, (v) pre- and post-contract obligations, (vi) transaction costs, (vii) inadequate financing in projects and (viii) high switching cost.

To comprehend the relationship among these variables, it is essential to model these factors for which an interpretive structural model has been developed. The details of the model so developed are discussed in the upcoming paragraphs.

2.4.1 Qualitative structural model

Following the steps presented in Figure 2.1, the ISM model has been developed. Based on the contextual association among various dimensions, the SSI matrix is developed. To arrive at a consensus, the SSI matrix so developed is discussed with experts from industry and academia. The SSI matrix was finalized based upon the responses and is presented in Table 2.8.

FRM is obtained from the initial RM. The driving power and dependency of all eight opportunistic behaviour risk factors of offshore outsourcing is presented in Table 2.9.

After creation of the FRM, further processing is done to attain the hierarchical model based on association among the dimensions. To do this, the sets of reachability and antecedent for all the dimensions are attained and their associated levels are found. The dimensions for which the sets of reachability and intersection are similar are placed at the top of ISM model.

In this case, the level identification is completed using five iterations for the eight dimensions of risk due to opportunistic behaviour risk

Table 2.8 SSI matrix of opportunistic behaviour risk dimensions

S. No.	Opportunistic behaviour risk dimensions	2	3	4	5	6	7	8
I	Miscommunication	O	V	V	V	O	O	O
2	Topographical distances		O	O	O	V	O	O
3	Partial work specifications			V	V	V	V	O
4	Span of control				O	O	V	V
5	Pre- and post-contract obligations					O	X	V
6	Transaction costs						O	O
7	Inadequate financing in project							V
8	High switching cost							

Table 2.9 Final reachability matrix of opportunistic behaviour risk dimensions

S. No.	Opportunistic behaviour risk factors	I	2	3	4	5	6	7	8	Driving power
I	Communication gap	I	0	I	I	I	I	I	I	7
2	Topographical distances	0	I	0	0	0	I	0	0	2
3	Partial work specifications	0	0	I	I	I	I	I	I	6
4	Span of control	0	0	0	I	I	0	I	I	4
5	Pre- and post-contract obligations	0	0	0	0	I	0	I	I	3
6	Transaction costs	0	0	0	0	0	I	0	0	I
7	Underinvestment in project	0	0	0	0	I	0	I	I	3
8	High switching cost	0	0	0	0	0	0	0	I	I
	Dependency	I	I	2	3	5	4	5	6	

dimensions as shown in Table 2.10. The levels so identified are used to build the ISM model.

Based on the final result of iterations, an initial diagraph is obtained which includes transitivity links. After removal of the transitivity links, a diagraph is obtained as shown by Figure 2.8.

From the diagraph, the ISM model is obtained by replacement of the nodes on the diagraph with name of the risk dimensions related to differences in culture in outsourcing as presented in Figure 2.9.

2.4.2 Model analysis and validation

To classify the dimensions and check their driving and dependence power, the MICMAC analysis has been done. Also, it is used to validate the

Table 2.10 Iterations of opportunistic behaviour risk dimensions

Dimensions	Reachability set	Antecedent set	Intersection set	Level
Iteration I				
1	1,3,4,5,6,7,8	1	1	
2	2,6	2	2	
3	3,4,5,6,7,8	1,3	3	
4	4,5,7,8	1,3,4	4	
5	5,7,8	1,3,4,5,7	5,7	
6	6	1,2,3,6	6	I
7	5,7,8	1,3,4,5,7	5,7	
8	8	1,3,4,5,7,8	8	I
Iteration II				
1	1,3,4,5, 7	1	1	
2	2	2	2	II
3	3,4,5, 7	1,3	3	
4	4,5,7	1,3,4	4	
5	5,7	1,3,4,5,7	5,7	II
7	5,7	1,3,4,5,7	5,7	II
Iteration III				
1	1,3,4	1	1	
3	3,4	1,3	3	
4	4	1,3,4	4	III
Iteration IV				
1	1,3	1	1	
3	3	1,3	3	IV
Iteration V				
1	1	1	1	V

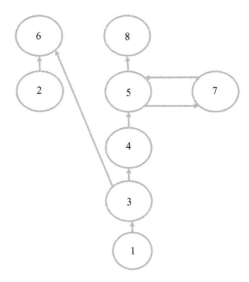

Figure 2.8 Diagraph of opportunistic behaviour risk dimensions.

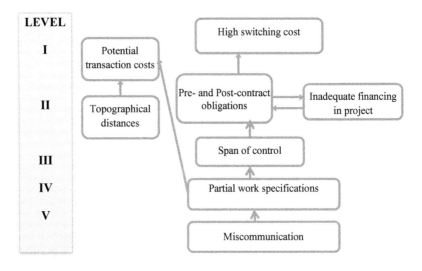

Figure 2.9 Qualitative model of opportunistic behaviour risk dimensions.

hierarchal model showing the dimensions of opportunistic behaviour risk factors in the study. Quadrants (i)–(iv) in Figure 2.10 show dimensions obtained from the MICMAC analysis.

On the basis of driver and dependency power, the opportunistic behaviour risk dimensions have been classified into four quadrants, viz. dependent, linkage, independent and autonomous, respectively. It is perceived that a risk dimension with high driving power is termed as a "key risk dimension"

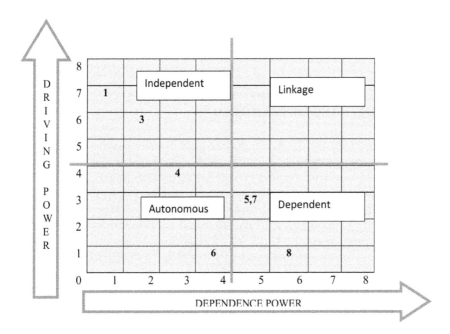

Figure 2.10 MICMAC analysis for opportunistic behaviour risk dimensions.

and it falls into the independent category or linkage category. It is observed that two risks, namely, communication gap and incomplete work specifications have high driving power and less dependence. If product specifications are partial or not fully clear in the beginning, a lot of continuous communication is required. Interpretation of work specifications is redefined as it suits the service provider due to incomplete work specifications.

It is also observed that post contractual behaviour, inadequate financing in project and switching cost all are weak drivers but they possess strong dependence on the other risk dimensions. Every relation has to be ended but in this case a lot of inside information remains with the service provider and it may be dangerous to the client organization. Redefine the downgraded task with new terminology and scale of measurement for achieving the target. For offshore outsourcing, not only training related to work is essential but they also should be aligned language wise and culture wise and huge costs are incurred due to switching the work from one service provider to another.

The MICMAC diagram (Figure 2.10) shows that geographical distances, level of control and potential transaction costs fall under autonomous category, thus they are not only weak as linkage but also weak as dependent dimensions too and hence do not possess much influence on the other risk dimensions of the system. Culture is also different due to topographical distances and it reflects on the behaviour and many times, the service provider takes advantage of this. In offshore outsourcing, the service provider organization has ample freedom and it may at certain times exhibit an opportunistic behaviour. Unwilling approach by employees of the client organization due to their layoff gives an upper hand to the service provider organization on specifications of the task.

2.5 INTELLECTUAL PROPERTY INFRINGEMENT RISK AND ITS DETERMINANTS

As discussed in Section 1.5.4, intellectual property risk is a major risk in offshore outsourcing business environment. Also, as per summarized views of respondents [Annexure Table A5] from selected business process outsourcing organizations, intellectual property risk may be due to (i) project size, (ii) research and development process, (iii) technical knowhow, (iv) lawless environment, (v) opportunistic behaviour and (vi) information technology.

In an attempt to comprehend the association among these dimensions, it is necessary to model these factors for which an interpretive structural model has been developed. The details of the model so developed are discussed in the upcoming paragraphs.

2.5.1 Qualitative structural model

Following the steps presented in Figure 2.1, the contextual association among various dimensions, the SSI matrix is developed. To arrive at a consensus, the SSI matrix so developed is discussed with experts from industry and academia. On the basis of responses, the SSI matrix was obtained and is presented in Table 2.11.

On replacing V, A, X and O by 1 and 0 based upon the association among dimensions, the SSI matrix is converted into a binary matrix. The binary matrix so obtained is called the initial RM. It is further converted into an FRM. Table 2.12 presents the powers (driving and dependence both) of the intellectual property risk dimensions.

After creation of the FRM, further processing is done to build the hierarchical model based on association among the dimensions. To do this, the sets of reachability and antecedent for each dimension are obtained for the

Table 2.11 SSI matrix of intellectual property risk dimensions

S. No.	Dimensions	2	3	4	5	6
1	Project size	O	O	A	A	A
2	Research and development process		O	A	A	O
3	Technical knowhow			O	O	X
4	Lawless environment				A	O
5	Opportunistic behaviour					V
6	Information technology					

Table 2.12 Reachability matrix of intellectual property risk dimensions

S. No.	Factor	1	2	3	4	5	6	Driving power
1	Project size	1	0	0	0	0	0	1
2	Research and development process	0	1	0	0	0	0	1
3	Technical knowhow	1	0	1	0	0	1	3
4	Lawless environment	1	1	0	1	0	0	3
5	Opportunistic behaviour	1	1	1	1	1	1	6
6	Information technology	1	0	1	0	0	1	3
	Dependence	5	3	3	2	1	3	

Table 2.13 Iterations of intellectual property infringement risk dimensions

Dimensions	Reachability set	Antecedent set	Intersection set	Level
Iteration I				
1	1	1,3,4,5,6	1	I
2	2	2,4,5	2	I
3	1,3,6	3,5,6	3	
4	1,2,4	4,5	4	
5	1,2,3,4,5,6	5	5	
6	1,3,6	3,5,6	3,6	
Iteration II				
3	3,6	3,5,6	3,6	II
4	4	4,5	4	II
5	3,4,5,6	5	5	
6	3,6	3,5,6	3,6	II
Iteration III				
5	5	5	5	III

identified dimensions and associated levels. The dimensions having reachability and similar intersection sets are placed at the top level in the ISM model. In this case, the level identification is completed using three iterations for the six dimensions of risk due to intellectual property. The three iterations are shown in Table 2.13.

After removal of the transitive links, the final diagraph is obtained (Figure 2.11) for intellectual property risk dimensions. If there is an association between the risk factor i with risk factor j as risk factor i influences risk factor j, then it is represented by an arrow pointing from risk factor i to risk factor j and the Arabic number within the circle of Figure 2.11 represents the risk factors.

On the basis of iterations, an initial diagraph is obtained which includes transitivity links. After removal of the transitive links, a diagraph in final form is obtained as presented by Figure 2.12.

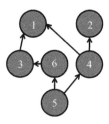

Figure 2.11 Diagraph of intellectual property infringement risk dimensions.

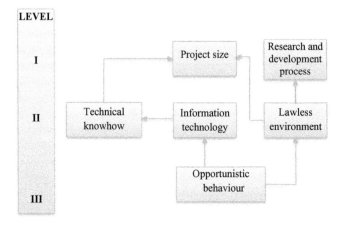

Figure 2.12 Qualitative model of intellectual property infringement risk dimensions.

2.5.2 Model analysis and validation

To classify the dimensions and check their driving and dependence power, the MICMAC analysis has been done. Also, it is used to validate the hierarchal model showing the dimensions of intellectual property risk factors in the study. Quadrants (i)–(iv) in Figure 2.13 show dimensions obtained from the MICMAC analysis.

It reveals that opportunistic behaviour has a strong driving power. It may be possible that a service provider organization uses the intellectual property of client organization to develop new things with collaboration of other organizations and renewal of the contract may be issued, to the service provider organization, due to threat of intellectual property loss.

Project size has strong dependence power and weak driving power. When the nature of work is intangible, the employee turnover of the service provider organization generates chances of intellectual property risk for the client organization. A group of persons from the existing project may emerge as competitors for the client organization for the same task by using their intellectual properties.

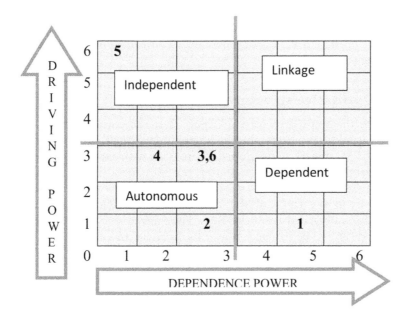

Figure 2.13 MICMAC diagram for intellectual property infringement risk dimensions.

Research and development process, technical knowhow, lawless environment and information technology all possess driver and dependence power weak. Offshore outsourcing of high-level work demands training at the initial stage about the overall scenario of the work and it imparts technical knowledge to the service provider organization. Chances of weak copyright and laws related to intellectual property exist in the service provider organization and these result in loss of intellectual property rights of the client organization. High-level coding is also decoded by the latest technological tool and it increases the vulnerability of important data related to intellectual property.

2.6 FINANCIAL RISK AND ITS DETERMINANTS

As discussed in Section 1.5.5, financial risk is one of the key risk dimensions among various offshore outsourcing risks which adversely affect the performance of the supply chain network. The sub-categories under financial risks related to offshore outsourcing as per summarized views of respondents from selected business process outsourcing organizations are (i) switching cost, (ii) transition and management cost, (iii) layoff cost, (iv) service provider selection cost, (v) measurement problems, (vi) costly contractual amendments, (vii) disputes and litigations and (viii) adaptation cost.

To comprehend the association among these dimensions, it is necessary to model these factors for which an interpretive structural model has been developed. The details of the model so developed are discussed in the upcoming paragraphs.

2.6.1 Qualitative structural model

Following the steps presented in Figure 2.1, the contextual association among various dimensions, the SSI matrix is developed. To arrive at a consensus, the SSI matrix so developed is discussed with experts from industry and academia. Based on the responses, the SSI matrix was finalized and is shown in Table 2.14.

Based upon the steps of the ISM procedure, FRM as presented in Table 2.15 is obtained from the RM. The matrix so obtained shows the driving power and dependence power for financial risk dimensions.

Table 2.14 SSIM of financial risk factors

S. No.	Factor	2	3	4	5	6	7	8
I	Switching cost	V	O	O	O	V	A	V
2	Transition and management cost		X	O	O	A	O	A
3	Layoff cost			O	O	O	A	O
4	Service provider selection cost				A	O	O	V
5	Measurement problems					V	V	V
6	Costly contractual amendments						A	V
7	Disputes and litigations							O
8	Adaptation cost							

Table 2.15 Final reachability matrix of financial risk dimensions

S. No.	Factor	I	2	3	4	5	6	7	8	Driving power
I	Switching cost	I	I	I	0	0	I	0	I	5
2	Transition and management cost	0	I	I	0	0	0	0	0	2
3	Layoff cost	0	I	I	0	0	0	0	0	2
4	Service provider selection cost	0	I	I	I	0	0	0	I	4
5	Measurement problems	0	I		I	I	0	0	I	5
6	Costly contractual amendments	0	I	I	0	0	I	0	I	4
7	Disputes and litigations	I	I	I	0	0	I	I	I	6
8	Adaptation cost	0	I	I	0	0	0	0	I	3
	Dependency	2	8	8	2	I	3	I	6	

After creation of the FRM, further processing is done to build the hierarchical model based on association among the dimensions. To do this, the sets of reachability and antecedent for each dimension are found for the identified dimensions and associated levels. The dimensions for which the sets of reachability and intersection are similar are placed at the top of the ISM model. In this case, the level identification is completed using four iterations for the eight dimensions of financial risk. The four iterations are shown in Table 2.16.

Based on the final result of iterations, an initial diagraph is obtained which includes transitivity links. After removal of the transitivity links, a diagraph is obtained as shown in Figure 2.14. From the diagraph, the ISM model is obtained by replacement of the nodes on the diagraph with the name of the risk dimensions related to financial risk in offshore outsourcing as presented in Figure 2.15.

Table 2.16 Iterations of financial risk factors

Dimension	Reachability set	Antecedent set	Intersection set	Level
Iteration I				
1	1,2,3,6,8	1,7	1	
2	2,3	1,2,3,4,5,6,7,8	2,3	I
3	2,3	1,2,3,4,5,6,7,8	2,3	I
4	2,3,4,8	4,5	4	
5	2,3,4,5,8	5	5	
6	2,3,6,8	1,6,7	6	
7	2,3,6,7,8	7	7	
8	2,3,8	1,4,5,6,7,8	8	
Iteration II				
1	1,6,8	1,7	1	
4	4,8	4,5	4	
5	4,5,8	5	5	
6	6,8	1,6,7	6	
7	6,7,8	7	7	
8	8	1,4,5,6,7,8	8	II
Iteration III				
1	1,6	1,7	1	
4	4	4,5	4	III
5	4,5	5	5	
6	6	1,6,7	6	III
7	6,7	7	7	
Iteration IV				
1	1	1,7	1	IV
5	5	5	5	IV
7	7	7	7	IV

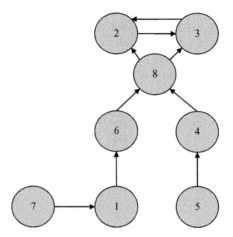

Figure 2.14 Diagraph of financial risk dimensions.

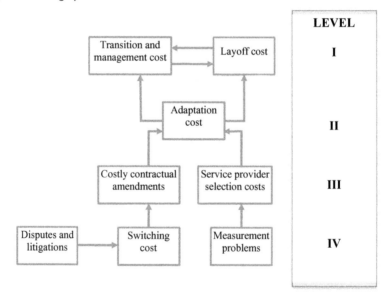

Figure 2.15 Qualitative model of financial risk dimensions.

2.6.2 Model analysis and validation

To classify the dimensions and check their driving and dependence power, the MICMAC analysis has been done. Also, it is used to validate the hierarchal model showing the dimensions of financial risk in the study. Quadrants (i)–(iv) in Figure 2.16 show dimensions obtained from the MICMAC analysis.

It is observed that three dimensions, namely, switching cost, disputes and litigations and measurement problems lie in the quadrant of independent

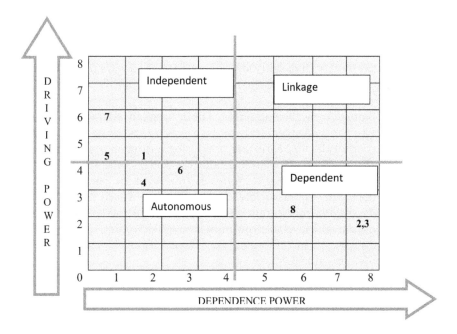

Figure 2.16 Driving power and dependence diagram for financial risk dimensions.

variables. It is observed that transition and management cost, layoff cost and adaptation cost come under the category of dependent variables. There is no dimension which lies in the linkage quadrant. It is observed that the service provider selection cost and costly contractual amendments fall in the category of an autonomous variable.

2.7 ORGANIZATION STRUCTURAL RISK AND ITS DETERMINANTS

As discussed in Section 1.5.6, the suppliers can alter their processes, change technologies and improve or change standard procedures without taking the clients into confidence and it results in organization structural risk as per summarized views of respondents [Annexure Table A7] from selected business process outsourcing organizations. This risk may be due to a number of determinants as (i) regulatory issues, (ii) incompatibility, (iii) delay in the completion of project, (iv) inappropriate resource allocation, (v) deficiency in capabilities and (vi) reduction in human capital.

To comprehend the relationship among these variables, it is essential to model these factors for which an interpretive structural model has been developed. The details of the model so developed are discussed in the upcoming paragraphs.

2.7.1 Qualitative structural model

Following the steps presented in Figure 2.1, the contextual association among various dimensions, the SSI matrix is developed. To arrive at a consensus, the SSI matrix so developed is discussed with experts from industry and academia. On the basis of the responses, the SSI matrix was finalized and is shown in Table 2.17.

FRM is obtained from the initial RM. The driving power and dependency power of organizational structural risk factors is presented in Table 2.18.

After creation of the FRM, further processing is done to obtain the hierarchical model based on association among the dimensions. To do this, the reachability set and antecedent set for each dimension are obtained for the identified dimensions and associated levels. The dimensions having reachability and the similar intersection sets are placed at the top level in the ISM model. In this case, the level identification is completed using four iterations for the six dimensions of the organization structural risk. The four iterations are shown in Table 2.19.

Based on the final result of iterations, an initial diagraph is obtained which includes transitivity links. After removing the transitivity links, a diagraph is obtained as shown by Figure 2.17. From the diagraph, structural model is obtained by replacement of the nodes on the diagraph with name of the risk dimensions related to organization structural risk dimensions as shown in Figure 2.18.

Table 2.17 SSIM of organization structural risk dimensions

S. No.	Dimensions	2	3	4	5	6
I	Regulatory issues	O	V	O	A	O
2	Incompatibility		V	O	X	A
3	Delay in completion of the project			A	A	A
4	Inappropriate resource allocation				V	X
5	Deficiency in capabilities					A
6	Reduction in human capital					

Table 2.18 Final reachability matrix of organization structural risk dimensions

S. No.	Dimensions	I	2	3	4	5	6	Driving power
I	Regulatory issues	I	0	I	0	0	0	2
2	Incompatibility	I	I	I	0	I	0	4
3	Delay in completion of the project	0	0	I	0	0	0	I
4	Inappropriate resource allocation	I	I	I	I	I	I	6
5	Deficiency in capabilities	I	I	I	0	I	0	4
6	Reduction in human capital	I	I	I	I	I	I	6
	Dependence	5	4	6	2	4	2	

Table 2.19 Iterations of organization structural risk dimensions

Dimension	Reachability set	Antecedent set	Intersection set	Level
Iteration I				
1	1,3	1,2,4,5,6	1	
2	1,2,3,5	2,4,5,6	2,5	
3	3	1,2,3,4,5,6	3	I
4	1,2,3,4,5,6	4,6	4,6	
5	1,2,3,5	2,4,5,6	2,5	
6	1,2,3,4,5,6	4,6	4,6	
Iteration II				
1	1	1,2,4,5,6	1	II
2	1,2,5	2,4,5,6	2,5	
4	1,2,4,5,6	4,6	4,6	
5	1,2,5	2,4,5,6	2,5	
6	1,2,4,5,6	4,6	4,6	
Iteration III				
2	2,5	2,4,5,6	2,5	III
4	2,4,5,6	4,6	4,6	
5	2,5	2,4,5,6	2,5	III
6	2,4,5,6	4,6	4,6	
Iteration IV				
4	4,6	4,6	4,6	IV
6	4,6	4,6	4,6	IV

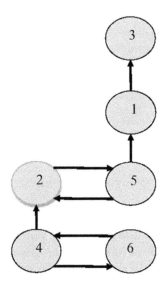

Figure 2.17 Diagraph of organization structural risk dimensions.

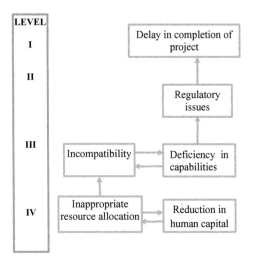

Figure 2.18 Qualitative model of organization structural risk dimensions.

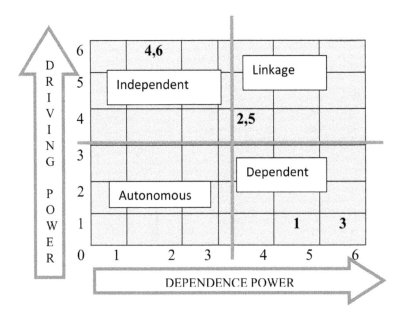

Figure 2.19 MICMAC diagram for organization structural risk dimensions.

2.7.2 Model analysis and validation

To classify the dimensions and check their powers of driving and dependence, the MICMAC analysis has been done. Also, it is used to validate the hierarchal model showing the dimensions of organization structural risk in the study. Quadrants (i)–(iv) in Figure 2.19 show dimensions obtained from the MICMAC analysis.

It is observed that two risk factors from the MICMAC diagram, namely, reduction in human capital and inappropriate resource allocation possess power high power of driving and low power of dependency.

It is also observed that delay in completion of project and regulatory issues are in the category of dependent factors as they possess weak driving power and strong dependency power on other dimensions and it is observed that deficient capabilities and incompatibility are in the category of linkage as they are strong drivers as well as strongly dependent on the other risk factors.

2.8 OPERATIONAL RISK AND ITS DETERMINANTS

As discussed in Section 1.5.7, many times defined targets are not fulfilled by outsourcing the service provider and there is a risk of being incorrect, not accomplishing what was in contract which is detrimental for offshore outsourcing business environment and also as per summarized views of respondents [Annexure Table A8] from selected business process outsourcing organizations. This risk may be due to a number of determinants like (i) poor delivery performance, (ii) lack of competency to fulfil a task, (iii) process fragmentation, (iv) conflict of objectives, (v) staff turnover, (vi) liabilities and litigations and (vii) poor service quality.

To comprehend the association among these dimensions, it is essential to model these factors for which an interpretive structural model has been developed. The details of the model so developed are discussed in the upcoming paragraphs.

2.8.1 Qualitative structural model

Following the steps presented in Figure 2.1, the contextual association among various dimensions, the SSI matrix is developed. To arrive at a consensus, the SSI matrix so developed is discussed with experts from industry and academia. On the basis of the responses, the SSI matrix was constructed and is shown in Table 2.20.

Based upon the steps of the ISM procedure, the RM is used to draw the FRM as shown in Table 2.21. The matrix so obtained shows the driving power and dependence for operational risk factors.

After creation of the FRM, further processing is done to obtain the hierarchical model based on association among the dimensions. To do this, the sets of reachability and antecedent for each dimension are found for the identified dimensions and associated levels. The dimensions having reachability and the similar intersection sets are placed at the top level in the ISM model. In this case, the level identification is completed using four iterations for the seven dimensions of operational risk. The four iterations are shown in Table 2.22.

Table 2.20 SSI matrix of operational risk dimensions

S. No.	Dimensions	2	3	4	5	6	7
I	Poor delivery performance	A	O	O	A	V	O
2	Lack of competency to fulfil task		O	A	A	O	V
3	Process fragmentation			A	O	V	V
4	Conflict of objectives				O	O	V
5	Staff turnover					O	V
6	Liabilities and litigations						A
7	Poor quality of service						

Table 2.21 Final reachability matrix of operational risk dimensions

S. No.	Dimensions	I	2	3	4	5	6	7	Driving power
I	Poor delivery performance	I	0	0	0	0	I	0	2
2	Lack of competency to fulfil task	I	I	0	0	0	I	I	4
3	Process fragmentation	0	0	I	0	0	I	I	3
4	Conflict of objectives	I	I	I	I	0	I	I	6
5	Staff turnover	I	I	0	0	I	I	I	5
6	Liabilities and litigations	0	0	0	0	0	I	0	I
7	Poor quality of service	0	0	0	0	0	I	I	2
	Dependence	4	3	2	I	I	7	5	

Table 2.22 First iteration of operational risk factors

Dimension	Reachability set	Antecedent set	Intersection set	Level
Iteration I				
I	1,6	1,2,3,4,5	I	
2	1,2,6,7	2,4,5	2	
3	3,6,7	3,4	3	
4	1,2,3,4,6,7	4	4	
5	1,2,5,6,7	5	5	
6	6	1,2,3,4,5,6,7	6	I
7	6,7	2,3,4,5,7	6	
Iteration II				
I	I	1,2,3,4,5	I	II
2	1,2,7	2,4,5	2	
3	3,7	3,4	3	
4	1,2,3,4,7	4	4	
5	1,2,5,7	5	5	
7	7	2,3,4,5,7	6	II

(Continued)

Table 2.22 (Continued) First iteration of operational risk factors

Dimension	Reachability set	Antecedent set	Intersection set	Level
Iteration III				
2	2	2,4,5	2	III
3	3	3,4	3	III
4	2,3,4	4	4	
5	2,5	5	5	
Iteration IV				
4	4	4	4	IV
5	5	5	5	IV

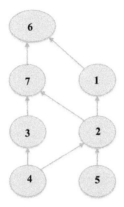

Figure 2.20 Diagraph of operational risk dimensions.

After removal of the transitive links, the final diagraph which is showing the inter-relationship among operational risk factors of offshore outsourcing is drawn in Figure 2.20 for operational risk dimensions. Arabic number within the circle of Figure 2.20 represents the risk dimensions.

The nodes of diagraph shown in Figure 2.20 are replaced with the name of operational risk dimensions and thus the diagraph is converted into an ISM structural model as presented in Figure 2.21.

2.8.2 Model analysis and validation

To classify the dimensions and check their driving and dependence power, the MICMAC analysis has been done. Also, it is used to validate the hierarchal model showing the dimensions of operational risk in the study. Quadrants (i)–(iv) in Figure 2.22 show dimensions obtained from the MICMAC analysis.

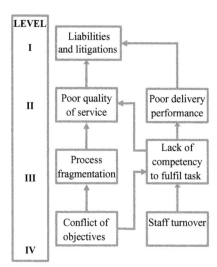

Figure 2.21 Qualitative model of operational risk dimensions.

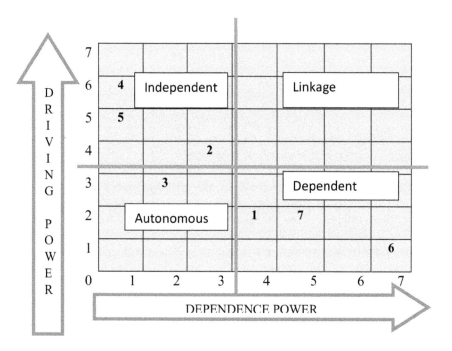

Figure 2.22 Driving power and dependence diagram for operational risk factors.

It is observed from the MICMAC diagram that three risk factors, namely, conflict of objectives, staff turnover and lack of competency to fulfil tasks possess high driving power and less power of dependency on other risk factors.

It is also observed that liabilities and litigations, poor service quality and poor delivery performance have strong dependence and weak driving power.

Figure 2.22 (MICMAC diagram) shows that process fragmentation is the only dimension which lies in the autonomous category.

2.9 CONCLUSION

According to Ishizaka et al. (2019), outsourcing business is associated with numerous risks which often lead to non-completion of projects in time. In addition, when outsourcing activity is accomplished through offshore vendors it results in risks in terms of language barriers, time zone, cultural differences and geographical barriers. This chapter develops qualitative structural models for the various types of offshore outsourcing risks like political risk, cultural risk due to diverse cultures, opportunistic behaviour risk, intellectual property infringement risk, financial risk, risk because of organizational structure and operational risk. The interpretive structural models are developed to understand the structural association among key dimensions in various offshore outsourcing risks and further, the use of the MICMAC analysis categorized the dimensions under four main categories: (i) independent dimensions, (ii) linkage dimensions, (iii) dependent dimensions and (iv) autonomous dimensions and validated the model obtained using an interpretive structural methodology in the study. The novelty of the ISM–MICMAC approach provides the potential researchers with a visual tool which helps to produce a universal viewpoint with respect to structural dependencies among outsourcing risks in a simple two-stage model which integrates the hierarchical structure model with the MICMAC analysis.

SUGGESTED READINGS

Aundhe, M.D., Mathew, S.K. (2009). Risks in offshore IT outsourcing: a service provider perspective. *European Management Journal*, 27, 418–428. https://doi.org/10.1016/j.emj.2009.01.004.

Babu, H., Bhardwaj, P., Agrawal, A.K. (2021). Modelling the supply chain risk variables using ISM: a case study on Indian manufacturing SMEs. *Journal of Modelling in Management*, 16(1), 215–239.

Chauhan, P., Kumar, S., Sharma, R.K. (2015). Qualitative and quantitative approach to model offshore outsourcing barriers due to cultural differences. *International Journal of Strategic Business Alliances*, 4(2/3), 184. https://doi.org/10.1504/ijsba.2015.072034.

Chauhan, P., Kumar, S., Sharma, R.K. (2017). Investigating the influence of opportunistic behaviour risk factors on offshore outsourcing. *International Journal of Business Excellence*, 12(2), 249. https://doi.org/10.1504/ijbex.2017.083570.

Christopher, M., Mena, C., Khan, O., Yurt, O. (2011). Approaches to managing global sourcing risk. *Supply Chain Management*, 16(2), 67–81. https://doi.org/10.1108/13598541111115338.

Gurtu, A., Johny, J. (2021). Supply chain risk management: literature review. *Risks*, 9, 16. https://doi.org/10.3390/risks9010016.

Lacity, M.C., Khan, S., Yan, A., Willcocks, L.P. (2010). A review of the IT outsourcing empirical literature and future research directions. *Journal of Information Technology*, 25(4), 395–433.

Nakano, M., Lau, A.K.W. (2020). A systematic review on supply chain risk management: using the strategy-structure-process-performance framework. *International Journal of Logistics Research and Applications*, 23(5), 443–473.

Patrucco, A.S., Scalera, V.G., Luzzini, D. (2016). Risks and governance modes in offshoring decisions: linking supply chain management and international business perspectives. *Supply Chain Forum: An International Journal*, 17(3), 170–182. https://doi.org/10.1080/16258312.2016.1219616.

Pisani, N., Ricart, J.E. (2016). Offshoring of services: a review of the literature and organizing framework. *Management International Review*, 56(3), 385–424.

Sharma, R.K., Chauhan, P. (2017). Investigating risks due to political environment as driver for other risks in offshore outsourcing. *International Journal of Business Performance and Supply Chain Modelling*, 9(2), 160. https://doi.org/10.1504/ijbpscm.2017.085493.

Sorooshian, S., Tavana, M., Ribeiro-Navarrete, S. (2023). From classical interpretive structural modeling to total interpretive structural modeling and beyond: a half-century of business research. *Journal of Business Research*, 157, 113642. https://doi.org/10.1016/j.jbusres.2022.113642.2. https://doi.org/10.1007/s40171-012-0008-3.

Szuster, M. (2013). Outsourcing and Offshoring as Factors Increasing Risk in Supply Chains. In: Szymczak, M. (eds) *Managing Towards Supply Chain Maturity*. Palgrave Macmillan, London. https://doi.org/10.1057/9781137359667_5.

Tate, W. L., Ellram, L. M., Bals, L., Hartmann, E. (2009). Offshore outsourcing of services: an evolutionary perspective. *International Journal of Production Economics*, 120, 512–524. https://doi.org/10.1016/j.ijpe.2009.04.005.

Chapter 3

Systemic qualitative model for offshore outsourcing

3.1 INTRODUCTION

Today in business offshore outsourcing has become a way of life and organizations are increasingly using outsourcing to streamline workflow and
save costs (BY SITEL STAFF OCTOBER 21, 2021). Because of the multifaceted nature of offshoring services and its linkage with cultures, disciplines and technologies, it has become important to analyse the association
among the various risks. After developing individual structural qualitative models for various offshore outsourcing risks in Chapter 2, this chapter presents an integrated model by taking into account the various risk
dimensions, namely, political risk, risk due to difference in culture, opportunistic behaviour risk, intellectual property risk, economic risk, organization structural risk and operational risk, compliance and regulatory
risk and lose control over talented information technology (IT) professionals. Qualitative analysis has been done through structured interviews
with respondents from various organizations and further, using SAP-LAP
(Situation, Actor, Processes; Learning, Action, Performance) analysis. An
interpretive structural modeling (ISM) approach is used to develop the
unified hierarchal offshore outsourcing model. The model groups various
important offshore outsourcing risk dimensions. At last, Matrice d'Impacts
Croisés Multiplication Applied to a Classification (MICMAC) analysis is
presented to validate the ISM model and analyse the risks with respect to
driver-dependence power. This analysis shall be beneficial for practising
supply chain managers in identifying and classifying essential criteria to
mitigate the risks.

3.2 QUALITATIVE ANALYSIS

To have a systemic view regarding various offshore outsourcing risks which
considerably affects the supply chain performance of companies, qualitative analysis has been conducted using a six-step procedure of SAP-LAP
shown in Figure 3.1. The semi-structured interviews have been conducted

DOI: 10.1201/9781032707884-3

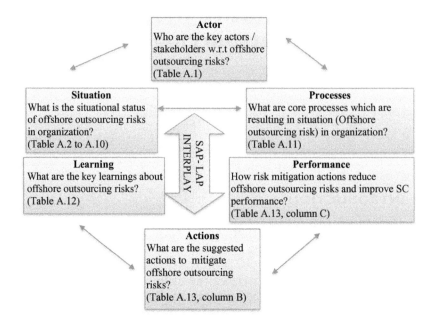

Figure 3.1 SAP-LAP framework for offshore outsourcing risks.

through the interactions with 25 executives of companies dealing with business process outsourcing. These companies mainly deal with business process service, research and development solutions, information technology service, financial service, analytics service, inventory service management, etc. Table A.1 presents the details of the respondent companies. The assessment survey on various offshore outsourcing risks using a questionnaire was planned using the existing literature studies. Responses from experts which include both academicians and industry practitioners dealing with offshore outsourcing were used to study the status of risk dimensions. Each respondent was asked to provide the response and assess the survey instrument for the understanding, readability and appropriateness. Summarized views with respect to nine offshore outsourcing risks within or across the organizations are presented in Tables A.2–A.10.

The views of different stakeholders on various offshore outsourcing risks through structured interviews were summarized, and it was found that with respect to client organizations there are certain issues related to offshore outsourcing risks which may not affect only one organization but also the other organization and vice versa. For instance, in the case of intellectual property risks, "the chance that confidential information may be leaked which may be the sole property of client organization". Similarly, the effects of other offshore outsourcing risks can be seen on various organizations collectively or independently.

3.3 SAP-LAP ANALYSIS OF OFFSHORE OUTSOURCING RISKS

To perform SAP-LAP, semi-structured interviews have been planned with respondents of firms engaged with outsourcing activities. Figure 3.1 presents the details of SAP-LAP analysis.

- *Actors:* The key actors/stakeholders with respect to offshore outsourcing risks include technology service providers, financial service providers, analytics service providers, business process service providers and providers of research and development solutions. Table A.1 provides the details of all service providers.
- *Situation:* The situational status of offshore outsourcing risks in respondent organizations is summarized and is represented from Tables A.2 to A.11 for instance, under risk due to cultural differences as quoted by stakeholders [R3], [R5], [R7], [R8], [R12], [R14] and [R20]. (Table A.3, point 9) *Due to different communication languages and cultures, members in one group may make distance from members of the other groups.*
- *Processes:* The core processes, viz. selecting a service provider organization, ability to move to standardized practices, standardization of IT processes, communication protocols, etc., are used to undertake offshore outsourcing activities successfully in the respective organizations. The 30 core processes are summarized and presented in Table A.12.
- *Learning:* The outcomes of key learning related to each risk in offshore outsourcing are tabulated category wise, as shown in Table A.13. For instance, it is shown that intellectual property risk has six dimensions, political risk has eight dimensions, etc.
- *Actions:* The actions to be undertaken by the organizations to address and mitigate the offshore outsourcing risks category wise are presented in column B of Table A.14. For instance, to mitigate intellectual property risk, the action to mitigate offshore outsourcing risk is to monitor opportunistic behaviour of the service provider organization. Similarly, to mitigate operational risk, the action is to resolve conflict of objectives between the organizations. All such actions are mentioned against the types of risks considered in the study.
- *Performances:* The effect of recommended actions on offshore outsourcing performance for mitigating offshore outsourcing risks is tabulated in column C of Table A.14. For instance, the recommended action to mitigate intellectual property risk is increased technical knowhow and enhanced competitive gains. The recommended action to mitigate the political risk is to build strong international relations and revamp fiscal and monetary policies.

On the basis of participant observant methodology and planned semi-structured interviews with stakeholders, SAP-LAP helps in identification and analysing of offshore outsourcing issues being faced by respondent firms. 4Ws and 1 H approach is used to investigate offshore outsourcing practice in respondent organizations. These 4 Ws and 1 H are: Who, Why, What, Where and How? For the stakeholders, it inquiries about their all-inclusive views, competences and roles with respect to the offshore outsourcing process.

Questions framed to conduct an interview are: *(i) What is the situational status of offshore outsourcing risks in an organization? (ii) Under which circumstances the offshoring outsourcing processes change? Mention them (iii) Why existing processes need to be modified or improved? and (iv) How do the various risk dimensions affect offshore outsourcing processes?*

Further, in the LAP analysis, "situation," "actor" and "process" categories are investigated in the form of key learnings about offshoring risks, actions required to mitigate the effect of risks and how these actions improve supply chain (SC) performance and mitigate risk.

3.4 SYSTEMIC STRUCTURAL MODEL FOR OFFSHORE OUTSOURCING RISKS

Following the steps presented in Figure 2.1 and discussed in Section 2.1, a systemic structural model has been developed. The paragraphs below present the step by step approach to construct the model.

3.4.1 Obtaining the contextual association among risk dimensions

Various risks in offshore outsourcing as discussed in Chapter 2 are considered for developing a systemic model of offshore outsourcing. These nine risks considered are risk due to political uncertainty, risk due to culture differences, opportunistic behaviour, intellectual property, economic risk, organization structural risk and operational risk, compliance and regulatory risk and lose control over talented IT professionals. For developing a model initially, a contextual association among risk dimensions is established. This means that how one risk affects the other? Experts from academic and industry were consulted to develop the casual relationship between the risk dimensions.

3.4.2 Building a structural self-interaction matrix

On the basis of the context association among the risk dimensions, a self-interaction matrix was constructed. The relationship among various risks and their factors are developed through a simple questionnaire after a brainstorming session so they are familiar with the context of survey and its implications.

It is filled by experts who are working as client account manager/outsourcing account delivery manager/service delivery manager/team manager/process manager/liaison officers/customer relationship executive/and one who looks after the offshore desk in the offshore outsourcing industry. With the help of extant literature, cases are cross-validated. Questionnaire is shown in the Appendix. Table 3.1 presents the details of self-interaction matrix.

3.4.3 Formulation of reachability matrix

The SSI matrix is changed into a 0,1 binary form of matrix as presented in Table 3.2, by replacing V, A, X and O symbols with 1 and 0. The conversion is based on the following arrangement:

- If (i, j) value in the SSI matrix is represented by V, then in the reachability matrix, (i, j) is represented by 1 and (j, i) is represented by 0.
- If (i, j) value in the SSI matrix is represented by A, then in the reachability matrix, (i, j) is represented by 0 and (j, i) is represented by 1.

Table 3.1 SSIM of offshore outsourcing risks

S. No.	Offshore outsourcing risk dimensions	2	3	4	5	6	7	8	9
1	Infringement of intellectual property	O	O	X	O	A	O	X	X
2	Regulatory and compliance risk		A	V	A	X	X	V	O
3	Political uncertainty risk			V	V	O	O	V	O
4	Operational failure risk				A	A	A	V	O
5	Cultural difference risk					O	V	V	O
6	Opportunistic behaviour risk						V	V	O
7	Organizational structure risk							V	O
8	Economic risk								O
9	Risk due to loss of talented IT professionals								

Table 3.2 IRM of offshore outsourcing risks

S. No.	Offshore outsourcing risk dimensions	1	2	3	4	5	6	7	8	9
1	Infringement of intellectual property	1	0	0	1	0	0	0	1	1
2	Regulatory and compliance risk	0	1	0	1	0	1	1	1	0
3	Political uncertainty risk	0	1	1	1	1	0	0	1	0
4	Operational failure risk	1	0	0	1	0	0	0	1	0
5	Cultural difference risk	0	1	0	1	1	0	1	1	0
6	Opportunistic behaviour risk	1	1	0	1	0	1	1	1	0
7	Organizational structure risk	0	1	0	1	0	0	1	1	0
8	Economic risk	1	0	0	0	0	0	0	1	0
9	Risk due to loss of talented IT professionals	1	0	0	0	0	0	0	0	1

- If (i, j) value in the SSI matrix is represented by X, then in the reachability matrix, (i, j) and (j, i) both are represented by 1.
- If (i, j) value in the SSI matrix is represented by O, then in the reachability matrix, (i, j) and (j, i) both are represented by 0.

3.4.4 Obtaining a final reachability matrix

Table 3.3 presents the details of the reachability matrix obtained initially by considering the transitive relationship among dimensions. Table 3.4 shows the final reachability matrix obtained from an initial reachability matrix with driving and dependence powers.

3.4.5 Level partitioning

After creation of the reachability matrix (final), further processing is done to make the hierarchical model based on association among the

Table 3.3 Initial reachability matrix of offshore outsourcing risks

S. No.	Offshore outsourcing risk dimensions	1	2	3	4	5	6	7	8	9
1	Infringement of intellectual property	1	0	0	1	0	0	0	1	1
2	Regulatory and compliance risk	1	1	0	1	0	1	1	1	1
3	Political uncertainty risk	1	1	1	1	1	1	1	1	1
4	Operational failure risk	1	0	0	1	0	0	0	1	1
5	Cultural difference risk	1	1	0	1	1	1	1	1	1
6	Opportunistic behaviour risk	1	1	0	1	0	1	1	1	1
7	Organizational structure risk	1	1	0	1	0	1	1	1	1
8	Economic risk	1	0	0	1	0	0	0	1	1
9	Risk due to loss of talented IT professionals	1	0	0	1	0	0	0	1	1

Table 3.4 Final reachability matrix of offshore outsourcing risks

S. No.	Offshore outsourcing risk dimensions	1	2	3	4	5	6	7	8	9	Driving power
1	Infringement of intellectual property	1	0	0	1	0	0	0	1	1	4
2	Regulatory and compliance risk	1	1	0	1	0	1	1	1	1	7
3	Political uncertainty risk	1	1	1	1	1	1	1	1	1	9
4	Operational failure risk	1	0	0	1	0	0	0	1	1	4
5	Cultural difference risk	1	1	0	1	1	1	1	1	1	8
6	Opportunistic behaviour risk	1	1	0	1	0	1	1	1	1	7
7	Organizational structure risk	1	1	0	1	0	1	1	1	1	7
8	Economic risk	1	0	0	1	0	0	0	1	1	4
9	Risk due to loss of talented IT professionals	1	0	0	1	0	0	0	1	1	4
		9	5	1	9	2	5	5	9	9	

dimensions. To do this, the sets of reachability and antecedent for each dimension are obtained for the identified dimensions and associated levels. The dimensions for which both sets of reachability and intersection are the same are positioned at the topmost position in the ISM model. The process is repeated until all levels of the structural model are determined. In this case, the level identification is completed using four iterations for the nine dimensions of offshore outsourcing. All these four iterations are as shown in Table 3.5.

3.4.6 Building of the ISM model

Based on the final result of iterations, an initial diagraph is obtained which includes transitivity links. After eliminating the transitivity links, a diagraph is drawn as shown by Figure 3.2. Arrow pointing from dimension i to dimension j shows the linkage among the risk dimensions in Figure 3.3.

Table 3.5 Iterations of offshore outsourcing risks

S. No.	Reachability set	Antecedent set	Intersection set	Level
Iteration I				
2	1,2,4,6,7,8,9	2,3,5,6,7	2,6,7	
3	1,2,3,4,5,6,7,8,9	3	3	
4	1,4,8,9	1,2,3,4,5,6,7,8,9	1,4,8,9	I
5	1,4,5,6,7,8,9	3,5,	5	
6	1,2,4,6,7,8,9	2,3,5,6,7	2,6,7	
7	1,2,4,6,7,8,9	2,3,5,6,7	2,6,7	
8	1,4,8,9	1,2,3,4,5,6,7,8,9	1,4,8,9	I
9	1,4,8,9	1,2,3,4,5,6,7,8,9	1,4,8,9	I
Iteration II				
2	2	2,3,6	2	II
3	2,3,5,6,7	3	3	
5	5,6,7	3,5,6	5,6	
6	2,5,6,7	3,5,6	5,6	
7	7	3,5,6,7	7	II
Iteration III				
3	3,5,6	3	3	
5	5,6	3,5,6	5,6	III
6	5,6	3,5,6	5,6	III
Iteration IV				
3	3	3	3	IV

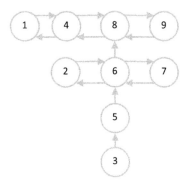

Figure 3.2 Diagraph of offshore outsourcing risks.

Figure 3.3 Qualitative structural model of offshore outsourcing risks with levels.

3.5 MODEL ANALYSIS AND VALIDATION

To assess the power of dimensions, viz., driver and dependent power and to further classify them under different clusters, namely, autonomous, linkage, dependent and Independent, MICMAC analysis as discussed in Section 2.1 is used. Figure 3.4 presents the results of MICMAC analysis which are used to validate the ISM for political risk.

1. *Autonomous risk dimensions:* Those dimensions possessing weakest powers of driving and dependence fall in the group of autonomous dimensions.
2. *Linkage risk dimensions:* The dimensions having the strongest powers of driving and dependence falls in the group of linkage dimensions.

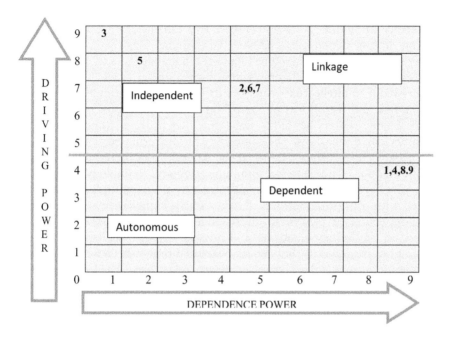

Figure 3.4 MICMAC diagram for offshore outsourcing risks a centre vertical line is to provided in figure after point 4 in horizontal axis.

3. *Dependent risk dimensions:* Dependent category contains dimensions which possess weak driving power and strong dependence power.
4. *Independent risk dimensions:* Here, in this category, dimensions possess strong power of driving and weak power of dependence.

3.6 CONCLUSION

Because of the multifaceted nature of offshoring services and its linkage with cultures, disciplines and technologies, it has become important to analyse the association among the various risks. In this chapter, a combined structural model is developed for the offshore outsourcing risk taking into account the various risk dimensions such as risk due to political uncertainty, risk due to differences in culture, opportunistic behaviour, infringement of intellectual property, economic risk, organizational structure, operational failure risk, compliance and regulatory risk and loss of talented IT professionals. Qualitative analysis has been done through semi-structured interviews with respondents from various organizations and further using SAP-LAP analysis. Finally, the MICMAC analysis is presented to validate the ISM model and examine the power of driver and dependent risk dimensions which may assist offshore managers in identification of various strategies to mitigate their undesirable effects.

From the MICMAC diagram shown in Figure 3.4, it is found that risk dimensions which exhibit strong driving and less dependence power on other risk dimensions are risk due to political uncertainty and risk due to difference in culture. As a result, these two dimensions are considered as strong driving dimensions and shall be thought of as core causes for all other risk dimensions, therefore supply chain managers must address them on priority for successful completion of offshore activities. It is perceived that risk due to intellectual property infringement, operational failure, economic risk and risk of loss of control on talented IT professionals possess weak driving powers but at the same time they possess strong dependence on other risk dimensions. The complexity of offshoring business processes, the geographical distance between the parties, the differences in culture, the limitations of information communication and technology (ICT) infrastructure between the two parties also enhance the magnitude of risks associated with offshoring business. Risks, namely, compliance and regulatory risk, opportunistic behaviour and risk due to organizational structure are placed in the linkage quadrant and possess stronger driving and dependence powers, and thus they should be treated timely. Because of increased competition among the businesses, firms today strive hard to remain in business by introducing new product capabilities and flexibilities thereof through offshore outsourcing by engagement of skilled professionals and offshoring certain activities. But these strategies comprise risks of opportunistic behaviour and may result in loss of implicit knowledge in the form of IPR infringements and related costs of disputes and litigations.

SUGGESTED READINGS

Arshinder, K.A., Deshmukh, S.G. (2007). Supply chain coordination issues: an SAP-LAP framework. *Asia Pacific Journal of Marketing and Logistics*, 19(3), 240–264. https://doi.org/10.1108/13555850710772923.

Aundhe, M.D., Mathew, S.K. (2009). Risks in offshore IT outsourcing: a service provider perspective. *European Management Journal*, 27, 418–428. https://doi.org/10.1016/j.emj.2009.01.004.

Christopher, M., Mena, C., Khan, O., Yurt, O. (2011). Approaches to managing global sourcing risk. *Supply Chain Management*, 16(2), 67–81. https://doi.org/10.1108/13598541111115338.

Gurtu, A., Johny, J. (2021). Supply chain risk management: literature review. *Risks*, 9, 16. https://doi.org/10.3390/risks9010016.

John, L., Ramesh, A. (2012). Humanitarian supply chain management in India: a SAP-LAP framework. *Journal of Advances in Management Research*, 9(2), 217–235. https://doi.org/10.1108/09727981211271968.

Patrucco, A.S., Scalera, V.G., Luzzini, D. (2016). Risks and governance modes in offshoring decisions: linking supply chain management and international business perspectives. Supply Chain Forum: *An International Journal*, 17(3), 170–182. https://doi.org/10.1080/16258312.2016.1219616.

Szuster, M. (2013). Outsourcing and Offshoring as Factors Increasing Risk in Supply Chains. In: Szymczak, M. (eds) *Managing Towards Supply Chain Maturity*. Palgrave Macmillan, London. https://doi.org/10.1057/9781137359667_5.

Chapter 4

Selecting vendor in offshore outsourcing environment

4.1 INTRODUCTION

Supply chain is defined as a linkage of various entities that are entrusted with the responsibilities of product development, material procurement and material movement between facilities, and manufacturing and distribution of finished products to the end customer. For a successful and competitive supply chain, each link must be very strong. The stronger the link, the easier and the better the process outcome; and the strength of the link in the supply chain principally depends on the vendor selection. Vendors play an important role in meeting the requirements of production, delivery and service with respect to the products. For any production or service industry, selection of the right vendors in the supply chain is a key success factor that reduces cost, increases customer satisfaction and improves competitiveness.

Selection of a vendor in an offshore outsourcing environment is not a trivial task as selection involves numerous criteria, viz. quality of the product/service, product variety, on-time deliveries, performance history of vendor, its financial position, guarantee/warranty requirements, environmental regulations, etc. Since last 25 years, numerous approaches have been used to develop and address the problem of supplier selection and their evaluation. These methods are broadly grouped into three clusters, viz. (i) multi-criteria decision-making, (ii) analytical programming models and (iii) heuristics-based approaches. The key criteria that affect the selection of a vendor in offshore outsourcing are its reliability, fiscal stability and capability to deliver the services within the assigned timeframe. For the success of offshore outsourcing, a firm must build a strong partnership with its suppliers. Thus, a decision-maker must consider the selection of the right supplier while designing a supply chain. The chapter presents a hybrid approach for selecting a vendor in which a conceptual framework has been built consisting of three stages. In stage 1, criteria related to vendor selection were identified from literature studies. Using these criteria a structural relationship model is built using an interpretive structural modeling (ISM)

DOI: 10.1201/9781032707884-4

approach. In stage 2, the dynamics of their driving-dependence power is examined using the Matrice d'Impacts Croisés Multiplication Applied to a Classification (MICMAC) method. In stage 3, a general fuzzy logic–based model is proposed to handle the problem of vendor selection. A case study based on the effect of cost, quality and service criteria on supplier selection is studied using the proposed fuzzy logic–based analysis by considering three probable suppliers from apparel manufacturing.

4.2 CRITERIA FOR VENDOR SELECTION

Vendor selection and management is the most difficult work in supply chain logistics and is termed in literature studies as the most challenging aspect. The various criteria used to select and manage vendors are supposed to be key strategic decisions to be undertaken by the management in a highly uncertain and competitive business domain. Hence, right vendor selection and management permit the firms to leverage their core capabilities to further enhance their supply chain responsiveness. The various dimensions which may have considerable influence on vendor selection (Chan and Kumar 2007; Ayag and Samanlioglu 2016; Aouadni et al. 2019; Naqvi and Amin 2021; Fallahpour et al. 2021; Rezaei et al. 2021) in an offshore outsourcing are defined as follows:

1. *Quality of product/service:* It is defined as the vendor's ability to meet the quality standards and desired specifications of products on a continual basis with accuracy and reliability.
2. *Service responsiveness:* It is stated as timeliness capability of the vendor which enables them quick deliveries or service schedules.
3. *Manufacturing flexibility:* It defines flexibility with respect to vendor's facilities, their manufacturing capabilities and their utilization rate.
4. *Cost:* It is defined as net price of the product or service after discount.
5. *Financial condition:* It represents the vendors' financial condition in terms of debt to equity ratio and liabilities.
6. *Technology:* Presence of skilled workforce with the state of art technology.
7. *Organizational management:* It is concerned with firms' ways of organization, management, coordination and utilization of resources effectively.
8. *Network design:* It is refers to the supply chain network.
9. *Trustworthiness:* It refers to the trust among the parties to complete the contract.
10. *Vendor relationship management*: It refers to the nature of vendor relations and dispute mitigation

11. *Regulatory compliance:* It refers to the ability of vendors to adhere to laws and requirements set forth by the government in sourcing and manufacturing of products.
12. *Managing risk and uncertainty:* It means capability of vendors to manage risk disruptions and ensure steadiness in services.

4.3 CONCEPTUAL FRAMEWORK FOR SUPPLIER SELECTION

The various criteria used to select and manage vendors as discussed in Section 4.2 are supposed to be key strategic decisions to be undertaken by the management in highly uncertain and competitive business domains. Hence, a framework is required to be built that helps the management for right vendor selection and permit the firms to leverage its core capabilities to further enhance its supply chain responsiveness. The conceptual framework as shown in Figure 4.1 has been developed to determine the association among various criteria defined for vendor selection.

The framework consists of three phases.

PHASE 1
   In phase 1, the criteria for vendor selection have been identified by examining literature studies followed by conversation with

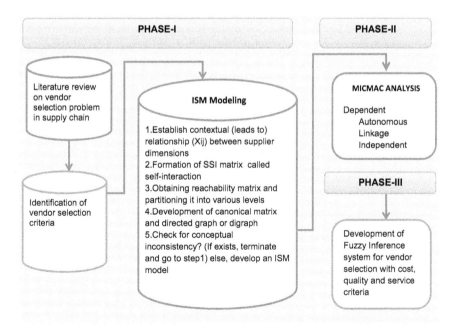

Figure 4.1 Conceptual framework.

professionals in the field of supply chain. Contemporary modeling approach based on the ISM approach is utilized to develop a model among these criteria. The use of the ISM approach is witnessed in literature studies by Rane and Kirkire (2017), Sharma and Sangal (2019), Sindhwani et al. (2022), Wankhade and Kundu (2020) and Hughes et al. (2020) in the area of medical devices, health services, software development, supply chain activities and engineering design problems.

PHASE 2

In phase 2, to study the dynamics of vendor selection criteria, MICMAC analysis is done based on driving and dependence power of criteria. It helps to classify the criteria into four quadrants, viz. dependent, independent, linkage and autonomous.

PHASE 3

In phase 3, a general fuzzy logic–based model is proposed to handle the vendor selection problem. A case study based on the effect of cost, quality and service criteria on supplier selection is studied by considering three probable suppliers from apparel manufacturing.

4.4 HIERARCHAL STRUCTURAL MODEL FOR SELECTING A VENDOR

4.4.1 Structural self-interaction matrix (SSIM)

The first step in the development of hierarchal structural model is construction of SSIM. This matrix is constructed on the basis of the contextual association among the vendor selection dimensions. The relationship among various dimensions is developed with the help of expert elicitation and through a simple questionnaire after a brainstorming session. It is filled by experts who are working as client account manager/outsourcing account delivery manager/service delivery manager/team manager/process manager/ liaison officers/customer relationship executive/and look after the offshore desk in the offshore outsourcing industry.

SSI matrix entries are presented in Table 4.1. The entries are in the form of symbols V, A, X and O, respectively.

4.4.2 Reachability matrix

The SSI matrix is changed into a 0, 1 binary form of matrix as presented in Table 3.2, by changing V, A, X and O with 1 and 0. The replacement is done on the basis of the below mentioned rules:

- If (i, j) value in the SSI matrix is represented by V, then (i, j) entry in the reachability matrix (RM) will be 1 and (j, i) entry will be 0.

Table 4.1 Matrix representing contextual association

Dimensions	Dimensions										
	12	*11*	*10*	*9*	*8*	*7*	*6*	*5*	*4*	*3*	*2*
1 **Quality of service/product**	O	O	A	O	O	A	A	O	V	A	O
2 **Service responsiveness**	O	A	A	A	A	A	O	O	O	A	
3 **Manufacturing flexibility**	O	A	O	V	O	O	O	O	V		
4 **Cost**	O	O	O	A	A	O	O	O			
5 **Economic factors**	O	V	O	V	O	O	O				
6 **Technology**	X	V	O	V	O	O					
7 **Organizational management**	V	O	V	V	O						
8 **Logistics and transportation network**	O	O	O	V							
9 **Trustworthiness**	V	V	A								
10 **Vendor relations**	O	O									
11 **Regulatory compliance**	A										
12 **Managing risk and uncertainty**											

- If (i, j) value in the SSI matrix is represented by A, then (i, j) entry in the RM will be 0 and (j, i) entry will be 1.
- If (i, j) value in the SSI matrix is represented by X, then both (i, j) and (j, i) entries in the RM will be 1.
- If (i, j) value in the SSI matrix is represented by O, both (i, j) and (j, i) entries in the RM will be 0.
- Based upon the above rules, the reachability matrix is obtained, as shown in Table 4.2.

From the initially prepared reachability matrix, transitivity is introduced to account for the indirect association among vendor dimensions. The * symbol in the final matrix of reachability is used to denote transitivity (Table 4.3).

4.4.3 Level partition

After creation of the final reachability matrix, further processing is done to make the hierarchical model based on association among the criteria. To do this, the sets of reachability and antecedent for each dimension are obtained for the identified dimensions and associated levels. The dimensions for which both sets of reachability and sets of intersection are the same are positioned at the top in the ISM hierarchy. The procedure is repeated until

Table 4.2 Reachability matrix

		Dimensions										
Dimensions	1	2	3	4	5	6	7	8	9	10	11	12
1 Quality of service/product	1	0	0	1	0	0	0	0	0	0	0	0
2 Service responsiveness	0	1	0	0	0	0	0	0	0	0	0	0
3 Manufacturing flexibility	1	1	1	1	0	0	0	0	1	0	0	0
4 Cost	0	0	0	1	0	0	0	0	0	0	0	0
5 Economic strength	0	0	0	0	1	0	0	0	1	0	1	0
6 Technology	1	0	0	0	0	1	0	0	1	0	1	1
7 Organizational management	1	1	0	0	0	0	1	0	1	1	0	1
8 Logistics and transportation network	0	1	0	1	0	0	0	1	1	0	0	0
9 Trustworthiness	0	1	0	1	0	0	0	0	1	0	1	1
10 Vendor relationship management	1	1	0	0	0	0	0	0	0	1	1	0
11 Regulatory compliance	0	1	1	0	0	0	0	0	0	0	1	0
12 Managing risk and uncertainty	0	0	0	0	0	1	0	0	0	0	1	1

Table 4.3 Final reachability matrix

						Criterion							Driving power
Criterion	1	2	3	4	5	6	7	8	9	10	11	12	
1 Quality of service/product	1	0	0	1	0	0	0	0	0	0	0	0	2
2 Service responsiveness	0	1	0	0	0	0	0	0	0	0	0	0	1
3 Manufacturing flexibility	1	1	1	1	0	1*	0	0	1	0	1*	1*	8
4 Cost	0	0	0	1	0	0	0	0	0	0	0	0	1
5 Economic factors	1*	1*	1*	1*	1	1*	0	0	1	0	1	1*	9
6 Technology	1	1*	1*	1*	0	1	0	0	1	0	1	1	8
7 Organizational management	1	1	1*	1*	0	1*	1	0	1	1	1*	1	10
8 Logistics and transportation	1*	1	1*	1	0	1*	0	1	1	0	1*	1*	9
9 Trustworthiness	1*	1	1*	1	0	1*	0	0	1	0	1	1	8
10 Vendor relations	1	1	1*	1*	0	1*	0	0	1	1	1*	1	9
11 Regulatory compliance	1*	1	1	1*	0	1*	0	0	1*	0	1	1*	8
12 Managing risk and uncertainties	1*	1*	1*	1*	0	1	0	0	1*	0	1	1	8
Dependence power	10	10	9	11	1	9	1	1	9	2	9	9	

all levels of the structural model are determined. In this case, the level identification is completed using five iterations for the twelve dimensions of vendor selection. The five iterations are shown in Table 4.4.

Table 4.4 Iterations

Criteria	Reachability	Antecedent	Intersection	Level
Iteration I				
1	1,4	1,3,5,6,7,8,9,10,11,12	1	
2	2	2,3,5,6,7,8,9,10,11,12	2	I
3	1,2,3,4,6,9,11,12	3,5,6,7,8,9,10,11,12	3,6,9,11,12	
4	4	1,3,4,5,6,7,8,9,10,11,12	4	I
5	1,2,3,4,5,6,9,11,12	5	5	
6	1,2,3,4,6,9,11,12	3,5,6,7,8,9,10,11,12	3,6,9,11,12	
7	1,2,3,4,6,7,9,10,11,12	7	7	
8	1,2,3,4,6,8,9,11,12	8	8	
9	1,2,3,4,6,9,11,12	3,5,6,7,8,9,10,11,12	3,6,9,11,12	
10	1,2,3,4,6,9,10,11,12	7,10	10	
11	1,2,3,4,6,9,11,12	3,5,6,7,8,9,10,11,12	3,6,9,11,12	
12	1,2,3,4,6,9,11,12	3,5,6,7,8,9,10,11,12	3,6,9,11,12	
Iteration II				
1	1	1,3,5,6,7,8,9,10,11,12	1	II
3	1,3,6,9,11,12	3,5,6,7,8,9,10,11,12	3,6,9,11,12	
5	1,3,5,6,9,11,12	5	5	
6	1,3,6,9,11,12	3,5,6,7,8,9,10,11,12	3,6,9,11,12	
7	1,3,6,7,9,10,11,12	7	7	
8	1,3,6,8,9,11,12	8	8	
9	1,3,6,9,11,12	3,5,6,7,8,9,10,11,12	3,6,9,11,12	
10	1,3,6,9,10,11,12	7,10	10	
11	1,3,6,9,11,12	3,5,6,7,8,9,10,11,12	3,6,9,11,12	
12	1,3,6,9,11,12	3,5,6,7,8,9,10,11,12	3,6,9,11,12	
Iteration III				
3	3,6,9,11,12	3,5,6,7,8,9,10,11,12	3,6,9,11,12	III
5	3,5,6,9,11,12	5	5	
6	3,6,9,11,12	3,5,6,7,8,9,10,11,12	3,6,9,11,12	III
7	3,6,7,9,10,11,12	7	7	
8	3,6,8,9,11,12	8	8	
9	3,6,9,11,12	3,5,6,7,8,9,10,11,12	3,6,9,11,12	III
10	3,6,9,10,11,12	7,10	10	
11	3,6,9,11,12	3,5,6,7,8,9,10,11,12	3,6,9,11,12	III
12	3,6,9,11,12	3,5,6,7,8,9,10,11,12	3,6,9,11,12	III

(Continued)

Table 4.4 (Continued) Iterations

Criteria	Reachability	Antecedent	Intersection	Level
Iteration IV				
5	5	5	5	IV
7	7,10	7	7	
8	8	8	8	IV
10	10	7,10	10	IV
Iteration V				
7	7	7	7	V

4.4.4 Formation of diagraph and ISM

Based on the final result of iterations, an initial diagraph is obtained which includes transitivity links. Figure 4.2 shows the ISM model for vendor selection. The vendor selection dimensions obtained in the first iteration are positioned at the top in the hierarchal model and the vendor selection dimensions obtained in the second iteration are positioned at the succeeding level and the process is continued till all criteria are placed in the hierarchal model.

4.5 MODEL ANALYSIS AND VALIDATION

To assess the power of dimensions, viz. driver and dependent power and to further classify them under different clusters, namely, autonomous, linkage, dependent and independent, MICMAC analysis as discussed in Section 2.1 is used. Figure 4.3 presents the results of MICMAC analysis which is used to validate the ISM model for vendor selection dimensions.

The four clusters with definitions are discussed as under:

1. *Autonomous dimensions:* Those dimensions possessing the weakest powers of driving and dependence fall in the group of autonomous dimensions.
2. *Linkage dimensions:* The dimensions having the strongest powers of driving and dependence falls in the group of linkage dimensions.
3. *Dependent risk dimensions:* This category contains criteria which have weak driving and strong dependence powers.
4. *Independent risk dimensions:* Here, in this category, dimensions possess strong power of driving and weak power of dependence.

The powers of driving and dependence for all the vendor criteria considered in the study are presented in Section 4.3. From Table 4.3, it is found that the driving power of criteria is shown in their respective rows with "1", and dependence power of criteria is indicated with column entries. With

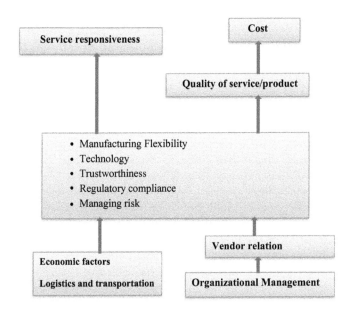

Figure 4.2 Interpretive structural modeling for vendor selection.

12												
11												
10	7		IV.					III. Linkage				
9	5,8	10	Independent									
8								3,6,9,				
								11,12				
7												
6												
5												
4			I. Autonomous					II. Dependent				
3												
2									1			
1									2	4		
	1	2	3	4	5	6	7	8	9	10	11	12

DRIVING POWER (left vertical label)

DEPENDENCE POWER

Figure 4.3 MICMAC diagram of criteria for vendor selection.

the help of the above facts, the driver-dependence diagram with four quadrants is made, as shown by Figure 4.3. For example, it can be concluded from Table 4.3 that vendor criterion 7, viz. organizational management possess score of 10 for driver power and score 1 for dependence. Hence, on the basis of scores, the criterion 7 is to be positioned in the fourth quadrant which is an independent cluster in the MICMAC Figure 4.3.

The following paragraphs discuss the four clusters and the criteria for vendor selection:

- *Quadrant I: Autonomous category*: No vendor selection criterion is found in this quadrant which states that all the criteria for vendor selection are important. Firms should consider all the criteria for selection of vendor in outsourcing the tasks or activities.
- *Quadrant II: Dependent category*: Among 12 criteria in the study, 3 criteria fall under this quadrant, viz. quality of service/product, service responsiveness and cost, respectively. All three criteria have less power of driving and high power of dependence. They are placed at the top position in the ISM hierarchy.
- *Quadrant III: Linkage category*: This category consists of five criteria, viz. manufacturing flexibility, technology, trustworthiness, regulatory compliance and managing risk. All five criteria are affected by low-level criteria in the model and hence they are called linkage criteria.
- *Quadrant IV: Independent category*: In this category, criteria, viz. economic environment, vendor relations, logistics and transportation and organizational management are found based upon the results of the ISM model. All four are known as independent dimensions with high driving and less dependence dynamics.

4.6 CASE STUDY ON VENDOR SELECTION

The section presents details of fuzzy logic for vendor selection with the help of a case study in which prospective vendor selection for considering textile and apparel manufacturing is considered. The function of supply chain in these firms is so difficult that it has turned out to be a challenge to manage vendor relations successfully. Thus, selecting the right vendors for raw material supplies is important to ensure that products are manufactured timely and are delivered as per the demand in right time, with right quality dimensions.

To achieve a reliable vendor performance, the selection depends on numerous criteria as discussed in Section 4.2. Numerous authors used cost, quality and service criteria in their approaches (Taherdoost et al. 2019; Aouadni et al. 2019; Naqvi and Amin 2021) to select probable vendors. In this section, quality of service, responsiveness and cost were used for the problem of vendor selection. There are sub-criteria to determine service or product quality, viz. less fault rate and acceptable process capabilities. For service responsiveness, the sub-criteria are on-time deliveries, flexibility and agility to respond to changes. For cost criterion, the sub-criteria considered are transportation mode, taxes, tariffs, etc. The appendix shows a sample questionnaire with a list of questions on a 1–5 point scale designed to gather important information from respondents working in textile and apparel firms.

The main modules of the proposed fuzzy system are:

- *Fuzzification inference unit*: This unit is used to define the numeric value of the inputs on the linguistic scale, i.e., low, high, medium, etc.
- *Knowledge base*: After fuzzifying the inputs and outputs using membership functions, the next step is generation of fuzzy rules in if-then format. The rules express the domain expert's knowledge. The knowledge can be gathered from qualified experts, databanks and previous literature. For example, if-then rule in fuzzy is composed as:
- *If X is A, THEN Y is B, where X and Y are linguistic variables and A and B are linguistic values determined by fuzzy sets on universe of discourse X and Y.*
- *Decision-making unit or inference unit*: It is the main unit of fuzzy-based decision support as it interprets the values in the input vector which are assigned to the output vector on the basis of pre-defined rules. In literature, the Mamdani-type and Sugeno-type are the two types of inference units used to handle such imprecise or vague information problems.
- *Defuzzification inference unit*: In this unit, conversion of a fuzzy number into a crisp output is done. The most commonly used defuzzification method is the centre of area method (COA), also known as the centroid method. This method determines the centre of area of fuzzy set and returns the corresponding crisp value.

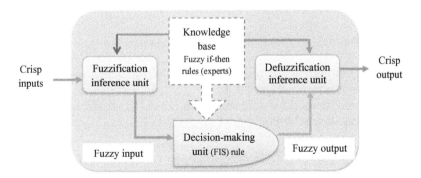

Figure 4.4 Fuzzy inference system.

Linguistic definitions: The inputs for vendor selection criteria, viz. quality rate, cost and service responsiveness are denoted by using linguistic terms, viz. low, medium and high.

Fuzzifying the inputs: For fuzzification of input criteria, trapezoidal membership functions are utilized, and for fuzzification of output criterion, viz. vendors (X, Y & Z), the triangular membership function is used.

The membership functions for the inputs in the model are presented in Figure 4.5, and the membership function for the output is presented in Figure 4.6. For the cost criterion, input range considered is 70%–100% based on works by Ghodsypour and O'Brien (2001) and Ali Ekici (2013) in which they advocated average product cost of 70%. The cost (material and service) component goes up to 80%. In the second criterion quality, the defect rate range considered is 0.01–0.09. The third criterion "service responsiveness" range is between 0.4 and 1 with three options as Good with the range of 0.40–0.55, better with the range of 0.50–0.85 and best with the range of 0.8–1.

If-then rules and inference system: This step provides a computational inference system based on if-then rules. On the basis of different combinations of input criteria for vendor selection, the if-then rule base is constructed. Jin (2000) in their work stated that fuzzy inference is based on various rules as much as feasible to fill up the input–output space of the inference system. On the contrary, it contains limited rules for the reason that a large number of rules affect the generality of the inference model. In the study, on the basis of 3 input criteria and terminology used to define these criteria, 27 rules are framed. The rules drafted in the fuzzy inference System (FIS) for the problem of vendor selection are presented in Figure 4.7. The proposed inference system performs the mapping of situation–action pairs based on input–output sets. Mamdani's inference system maps the input–output. Equation 4.1 shows the operator (max–min) used in the Mamdani inference system.

$$\mu B'(y) = \text{max}-\text{min}(\beta_k, \mu_{B_k}(y)), \tag{4.1}$$

where $\beta_k = \text{min}\,\alpha_{i,k}\left[\alpha_{i,k} = \text{sup min}\left(\mu'_A(x_i), \mu_{Ai,k}(x)\right)\right]$

Equation 4.1 is graphically presented in Figure 4.8. The inference process for two rules is given by Figure 4.8.

The centroid method for defuzzification is used to obtain aggregation as represented as shown by Equation 4.2:

$$y = \int y\mu\text{Aggr}(y)dy / \int \mu\text{Aggr}(y)dy \tag{4.2}$$

where numerator μAggr(y): fuzzy set after aggregation
$\int \mu$Aggr(y): is the area below μAggr(y)

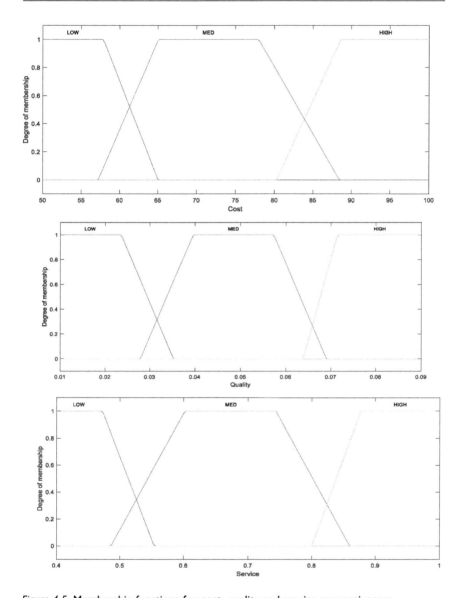

Figure 4.5 Membership functions for cost, quality and service responsiveness.

4.7 RESULTS AND DISCUSSIONS

As discussed in Section 4.5, the concept of fuzzy logic has been used to select vendors in offshore environment with respect to textile and apparel manufacturing firms. The subjective valuation on five-point scale using a questionnaire is used to gather the on-hand information related to sub-criteria

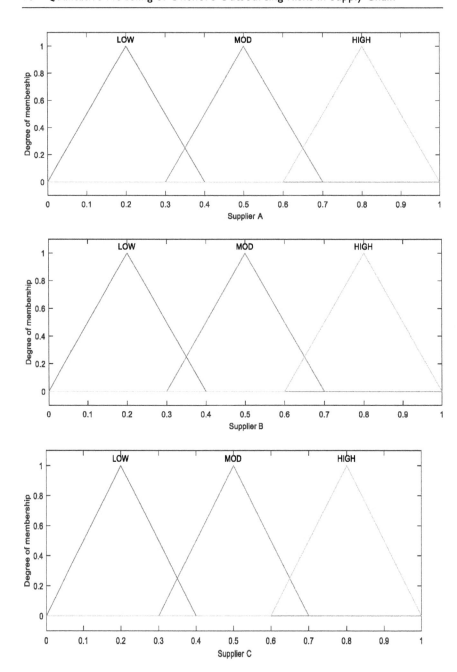

Figure 4.6 Membership functions for output

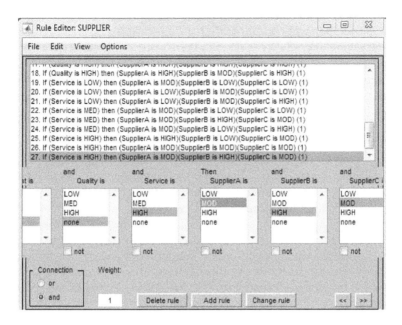

Figure 4.7 Rules drafted in the FIS editor.

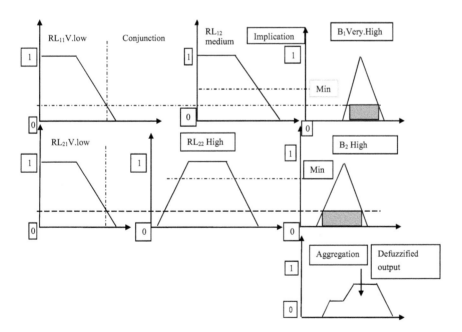

Figure 4.8 Fuzzy inference process in the Mamdani system.

signifying the quality of service/product and delivery or service rate. The supply chain managers accountable for decisions related to procurement, manufacturing and quality control in their firms were contacted to get the desired responses. Moreover, the knowledge and insight of supply chain professionals on quality of service/product and delivery or service rate forms the basis for performing simulation under various settings of criteria for vendor selection.

Experiment-I considers the quality of service/product – low, service responsiveness – high and cost – high. For these three criteria, the simulation output with respect to three vendors so found is 0.502 (Vendor X), 0.498 (Vendor Y) and 0.583 (Vendor Z), respectively. From these values, it is observed that the performance of Vendor Z is better followed by Vendor X and hence Vendor Z is recommended.

Experiment II considers the quality of service/product – moderate, service responsiveness – high and cost – moderate. For these three criteria, the simulation output with respect to three vendors so found is 0.54 (Vendor X), 0.460 (Vendor Y) and 0.444 (Vendor Z), respectively. From these values, it is observed that the performance of Vendor X is better and hence Vendor X is recommended.

Experiment III considers the quality of service/product – low, service responsiveness – moderate and cost – high. For these three criteria, the simulation output with respect to three vendors so found is 0.478 (Vendor X), 0.522 (Vendor Y) and 0.504 (Vendor Z), respectively. From these values, it is observed that performance of Vendor Y is better and hence Vendor Y is recommended.

More experimental settings can be worked out with inputs related to quality of service/product and delivery or service rate and based on the needs of the firms, vendor selection can be done.

Table 4.5 shows simulation results of experiments with various input criteria, viz. cost, quality of service/product and responsiveness or service rate. The output scores for Vendors X, Y and Z are presented in the table with recommendations. Figure 4.9a–c presents the simulation results. For more details, readers can refer the literature by Kumar and Pani (2014), Jain et al. (2018), Eydi and Fazli (2019) and Fallahpour et al. (2021) for vendor selection in fuzzy and uncertain environment based on various dimensions.

4.8 CONCLUSION

The chapter on vendor selection in offshore outsourcing discusses a conceptual approach to support supply chain managers in decision-making when they are encountered with the problem of vendor selection. A set of 12 criteria are identified from the literature which may have considerable influence on the problem of vendor selection. To analyse the structural

Table 4.5 Experiment settings with output and recommendations

	Input		Output scores			Final recommendation	
Simulation experiment 1	*Criteria*	*Linguistic definition*	*Vendor A*	*Vendor B*	*Vendor C*	*Vendor recommendation*	
	COST	High (92.6)	0.502	0.498	0.583	Vendor C	
	QUALITY	Low (0.027)					
	SERVICE RESPONSIVENESS	High (1)					
Simulation experiment 2	Criteria	Linguistic definition	Vendor A	Vendor B	Vendor C	Vendor recommendation	
	COST	Mod (77.1)	0.540	0.460	0.444	Vendor A	
	QUALITY	Mod (0.048)					
	SERVICE RESPONSIVENESS	High (1)					
Simulation experiment 3	Criteria	Linguistic definition	Vendor A	Vendor B	Vendor C	Vendor recommendation	
	COST	High (100)	0.478	0.522	0.504	Vendor B	
	QUALITY	Low (0.035)					
	SERVICE RESPONSIVENESS	Moderate (0.744)					

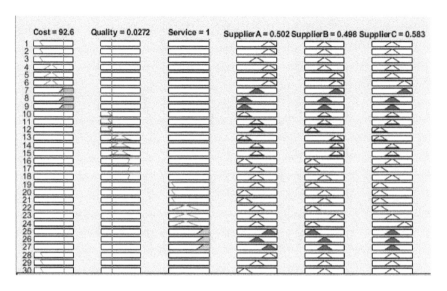

Figure 4.9 Simulation results for experimental setting I.

association among these criteria, an ISM is built. It represents these criteria at five significant levels. Level I presents the criterion organizational management followed by criteria, viz. economic factors, logistics and transportation and vendor relations at Level II. All these criteria are placed in quadrant IV of the MICMAC diagram showing strong dependence and weak driving power. Five criteria are placed at Level III of the ISM model. These are manufacturing flexibility, technology, trustworthiness, regulatory compliance and managing risk. All these are affected by low-level criteria in the model and hence they are called linkage criteria. Level IV in the ISM model portrays quality of service/product as criterion and the top Level V shows two criteria, i.e., delivery rate/service and cost respectively.

Furthermore, a case study has been presented to showcase the importance of dependent criteria, namely, quality, delivery service and cost in vendor selection. The case study makes use of fuzzy logic approach. FIS uses Mamdani logic to conduct experiments with different settings of the criteria. Data are collected from the firms engaged in textile and apparel manufacturing on various sub-criteria using the 1–5 point scale. The simulation results show the most suitable vendor under different settings of input criteria. For example, when the cost criterion selected is high, the quality criterion selected is low and service responsiveness selected is high, then performance of supplier Z comes better and hence recommendation is made for supplier Z. The results illustrate that the vendor selection may vary with the organizational needs and perceived level of criteria.

SUGGESTED READINGS

Aouadni, S., Aouadni, I., Rebai, A. (2019). A systematic review on supplier selection and order allocation problems. *Journal of Industrial Engineering International*, 15, 267–289.

Ayağ, Z., Samanlioglu, F. (2016). An intelligent approach to supplier evaluation in automotive sector. *Journal of Intelligent Manufacturing*, 27, 889–903.

Jain, V., Sangaiah, A. K., Sakhuja, S. (2018). Supplier selection using fuzzy AHP and TOPSIS: a case study in the Indian automotive industry. *Neural Computing and Applications*, 29, 555–564.

Kumar Kar, A., Pani, A. K. (2014). Exploring the importance of different supplier selection criteria. *Management Research Review*, 37(1), 89–105.

Luthra, S., Govindan, K., Kannan, D., Mangla, S. K., Garg, C. P. (2017). An integrated framework for sustainable supplier selection and evaluation in supply chains. *Journal of Cleaner Production*, 140, 1686–1698.

Mohammed, A., de Sousa Jabbour, L., Beatriz, A., Koh, L., Hubbard, N., Jabbour, C., Jose, C., Al Ahmed, T. (2022). The sourcing decision-making process in the era of digitalization: a new quantitative methodology. *Transportation Research Part E: Logistics and Transportation Review, Elsevier*, 168(C), 102948.

SevincIlhan, O. (2013). An intelligent supplier evaluation, selection and development system. *Applied Soft Computing*, 13(1), 690–697. https://doi.org/10.1016/j.asoc.2012.08.008.

Raut, R. D., Gardas, B. B., Narkhede, B. E., Zhang, L. L. (2020). Supplier selection and performance evaluation for formulating supplier selection strategy by MCDM-based approach. *International Journal of Business Excellence*, 20(4), 500–520.

Rezaei, A., Aghsami, A., Rabbani, M. (2021). Supplier selection and order allocation model with disruption and environmental risks in centralized supply chain. *International Journal of System Assurance Engineering and Management*, 12, 1036–1072. https://doi.org/10.1007/s13198-021-01164-1.

Sharma, R.K. (2022). Examining interaction among supplier selection strategies in an outsourcing environment using ISM and fuzzy logic approach. *International Journal of System Assurance Engineering and Management*, 13, 2175–2194. https://doi.org/10.1007/s13198-022-01624-2.

APPENDIX: QUESTIONNAIRE ON VENDOR SELECTION CRITERIA

Fraction Defective/No. of Defects

	1	2	3	4	5	
Least important	○	○	○	○	○	Highly important

Process capability

	1	2	3	4	5	
Least important	○	○	○	○	○	Highly important

Transportation cost

	1	2	3	4	5	
Least important	○	○	○	○	○	Highly important

Raw material cost

	1	2	3	4	5	
Least important	○	○	○	○	○	Highly important

Warehouse/storage

	1	2	3	4	5	
Least important	○	○	○	○	○	Highly important

Tariff and taxes

	1	2	3	4	5	
Least important	○	○	○	○	○	Highly important

Response to changes/ process flexibility

	1	2	3	4	5	
Least important	○	○	○	○	○	Highly important

Chapter 5

Perfect order fulfilment in supply chain logistics

5.1 INTRODUCTION

With increased globalization, the coordination among various entities of the supply chain (SC) has become difficult. As a result, delivering on-time and in full (OTIF) which is called perfect order fulfilment (POF) has become challenging as it is dependent on different criteria, viz. demand forecasting and its accuracy, capacity decisions, technological advancements, etc. Fulfilment of the orders plays a major role in the supply chain process. To increase supply chain efficiency, POF is an important SC dimension to study. It helps to measure the performance of supply chain in shipping the right product, at the right place, at the right time, in the right condition and right packaging, with the right quantity to the end customer. The following equation is used to represent the perfect order:

$$\text{Perfect Order} = f \begin{bmatrix} \text{Delivery Time } (DT), \text{Quantity Delivered } (Qty), \\ \text{Quality } (Qual) \end{bmatrix}$$

(5.1)

In addition, it needs the formation of a well-designed network consisting of all SC stakeholders and integration among them. Dimensions affecting order fulfilment from the supplier to the manufacturer's place are considered upstream dimensions and the dimensions which effect order management from distributor to the end customer are called as downstream dimensions.

With respect to its advantages, POF helps the businesses to improve their supply chain reliability by integrating the functions of all the entities in the supply chain network. It also helps to optimize the supply chain process and reduce stock-outs related to inventory, revenue losses, forecasting and transportation costs, and thus provide addition to the value of products.

In this chapter, various dimensions that affect POF in upstream and downstream supply chain networks are identified and the structural association among the dimensions is analysed using the interpretive structural modeling (ISM) approach. Furthermore, Matrice d'Impacts Croisés

DOI: 10.1201/9781032707884-5

Multiplication Applied to a Classification (MICMAC) analysis is used to understand the dynamics of POF dimensions in a multi-echelon supply chain. The knowledge of dimensions helps the managers to manage orders of their clients and make their supply chains more responsive and reliable. The next section presents fulfilment strategies being adopted by e-commerce companies.

5.2 PERFECT ORDER FULFILMENT STRATEGIES

Today, e-commerce companies enjoy competitive advantage over their traditional companies because of their ability to please the customer with low price tags, wider choices and smoother delivery schedules. This is because of the following strategies practiced by e-commerce companies for order fulfilment (Lakri et al. 2015; Bhattacharya and Chetty 2019) as discussed under:

Third-party (3P) order fulfilment: In this strategy, a product is obtained in bulk quantity from the manufacturing firm and then the product is shipped to the outsourced party for packaging and shipping to the client as per their orders.

Distributed delivery centre–based fulfilment: In this strategy, retailer firms in e-commerce business maintain numerous distribution centres operated by them. In this way, based upon the customer orders, fulfilment of orders is achieved through distribution centres.

Build to order: In such a strategy, the manufacturer produces the product based on demand as per the order received and confirmed from the end customer. It is the oldest style of fulfilling the orders and is used for extremely customized or low-volume products.

Fulfilment by supply chain partners: The order fulfilment process is disseminated between partners to regulate the persistent deliveries to the customers. The partnerships are done in both vertical and horizontal collaboration. They provide solutions for sourcing, quality assurance and logistical operations with the help of a dedicated network of numerous suppliers.

Drop shipping: Drop shipping is an order fulfilment strategy in which the retailer does not house products in their inventory, but they rely on wholesaler or manufacturer for shipment of orders to customers. The model has its advantages, which include lower capital investment, speed to market and scalability. Low margins, long shipping durations and issues with suppliers are some of the disadvantages of drop shipping. Amazon is a perfect example of an e-commerce retailer following the drop shipping strategy.

Traditional order fulfilment strategies: Traditional methods for order fulfilment are distribution centre–based order execution, warehousing and vendor-based direct fulfilment.

Distribution centre approach: The wholesaler regulates the stock of goods for customers in the centralized distribution centre in this approach. They're customer-centric, build to serve retail stores directly. Suppliers frequently ship goods to these centres, to serve definite retail stores at various locations. If as a customer you place an order for a product from Walmart online, and select to pick it up in-store, the nearest distribution centre will make sure the availability and delivery of the product. Today these centres are equipped with latest technology which ensures order processing and order delivery within the shortest period of time. Large and long-term storage with affordable costs, location near to market and quick delivery capability are few advantages of a distribution centre–based strategy.

Warehousing: Warehouse in the logistics system works as a centralized location for the storage of goods. It means that the warehouse is used to receive, store and then distribute products from one location to another. Thus, the most common functions of warehouse are purchase of goods, managing inventories, material handling, packaging and shipment of goods. Because of the numerous functions, warehouses are used by distributors, manufacturers, importers, exporters, businesses and wholesalers. Main categories of warehouses are public, private and bonded warehouses. Public warehouses owned either by individual or a corporation provide affordable space and thus they play an important role in the success of e-commerce start-up firms today. Private warehouses are managed by big corporations and manufacturers. Bonded warehouses are governed by government and are used to store imported goods, which are released after the payment of customs duty by the owner of the goods.

Vendor-based direct fulfilment: To out-shine their competitors, today e-commerce retailers are endeavouring to provide best services at the most competitive rates to their clients. Direct fulfilment is the process of distributing goods ordered by a customer directly, using firm's own resources, and not relying upon third-party logistics service providers for fulfilment process. For instance, fulfilment by Amazon is a service provided by Amazon that helps their customers with storing, packaging, distributing guidelines and other facets of direct fulfilment. This strategy reduces the load on vendors. Vendors send their products to an Amazon fulfilment centre, where storage of goods is done based on their demand and expiry dates, etc. The shipments are made as per the need. The main benefits of vendor-based fulfilment are the ability to gain first-hand look on the product performance and consolidate important information with respect to demand and supply of goods. It makes direct relationship with clients with hassle-free returns.

The next section presents the conceptual framework to model POF in the supply chain.

5.3 CONCEPTUAL FRAMEWORK FOR POF IN SUPPLY CHAIN LOGISTICS

Figure 5.1 shows the conceptual framework for POF in the supply chain. It consists of two stages. In the first stage, both the upstream and downstream dimensions which affect the POF in the supply chain network have been identified through literature studies (Sheel et al. 2020; Puska et al. 2020; Akhtar et al. 2021; Lakri et al. 2015; Najmi et al. 2013) and discussion with domain experts. In the second phase, the ISM method has been applied for modeling the structural association among these variables by following the steps as discussed in Section 2.1.

The various dimensions which affect order fulfilment in upstream supply chain are:

- Suppliers complexity
- Procurement source
- Flow complications
- Internationalization
- Coordination
- Technological advances
- Employee turnover rate
- Demand forecast
- Capacity
- Information sharing

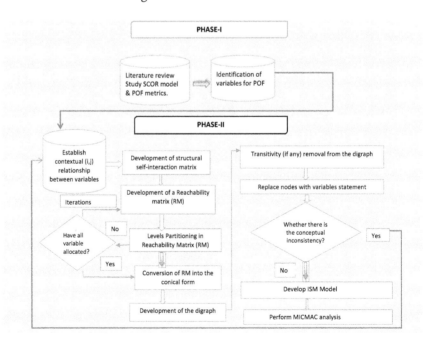

Figure 5.1 Conceptual framework.

Similarly, for downstream supply chain, the order fulfilment dimensions are:

- Lead time
- Precise data
- Responsiveness
- Inventory
- Product quality
- Product volume
- Number of echelons
- Communication and information sharing
- Client feedback
- Demand forecast
- Shipping duration

Tables 5.1 and 5.2 present the definitions of the dimensions listed above.

Table 5.1 Upstream dimensions for perfect order fulfilment process

S. No.	Dimensions	Definitions
1	Suppliers' complexity	It is defined as the number of suppliers/vendors levels (tier 1, tier 2, tier 3, etc.) in the upstream supply chain network who provides supplies to the manufacturer based upon the ordered quantity.
2	Capacity planning	It relates to the extent up to which an organization uses its installed capacity to get maximum profits.
3	Transportation	It defines the complexity associated with the flow of material in supply chain network consisting of multiple partners.
4	Internationalization	Internationalization is defined as a way in which firms in overseas countries interact with each other and integrate themselves for flow of information, goods and capital across their borders.
5	Supply chain integration	It is associated with linkage of the activities among supply chain entities for successful supply chain.
6	Technological advances	It defines the introduction and use of technology for increasing organizational productivity and efficient resource utilization.
7	Employee turnover rate	Employee turnover rate defines the percentage of workforce who leaves the company and is substituted by new employees.
8	Demand forecast	It is a process by which the historical sales data are used to obtain an estimation of future customer demand.
9	Procurement source	It is stated as a number of offshore/on-shore suppliers on whom the organizations depends for the supply or procurement of raw material.
10	Information sharing	It deals with sharing of information among the upstream supply chain entities in a supply chain network.

Table 5.2 Downstream dimensions for perfect order fulfilment process

S. No.	Dimensions	Definitions
1	Lead time	Lead time is defined as the timeline used to denote the overall duration or interval of processing the order until the customer receives it.
2	Precise data	It is related with the precision and accuracy of data managed by SC members. For least deviation in order fulfilment, data need to be precise.
3	Responsiveness	Ability to quickly adapt to the changes in demand.
4	Inventory	It is defined as quantity of stock with the firms for fulfilment of orders.
5	Product quality	Quality is the characteristic which defines the features of a product that clients wish to pay for their orders.
6	Product volume	It is defined as the number of the order/size the entities, i.e., retailers, wholesalers, distributors place in the supply chain network
7	Number of echelons	This defines the number of entities a supply chain network comprises. The more the number of echelons, the more the chances of order amplification.
8	Communication and information sharing	It deals with sharing of information among the entities in a supply chain network.
9	Customer queries/ feedbacks	It aims at making customer satisfied through their feedbacks/reviews and queries related to product functioning.
10	Demand forecast	Forecasting uses historical data to cope up with the uncertain business environment and forecast of future demand.
11	Shipping duration	The time taken by the firms to deliver the product to the client; the lesser the delivery time, the more satisfied the customer is.

5.4 HIERARCHAL STRUCTURAL MODEL FOR POF IN UPSTREAM SUPPLY CHAIN LOGISTICS

The model has been developed by following various steps as shown in Figure 5.1 and discussed in Section 5.3.

5.4.1 Structural self-interaction matrix (SSIM)

The relative association among the dimensions is represented as "lead to", which states that how one dimension leads to another dimension. Consultations with a total of six experts, out of which four were from the outsourcing companies and two were from academic units were made to formulate the contextual relation among the upstream dimensions. On the basis of responses, structured self-interaction (SSI) matrix is made as presented in Table 5.3. The matrix was obtained.

Table 5.3 Structured self-interaction matrix (SSIM)

Dimensions	1	2	3	4	5	6	7	8	9
Supplier complexity		O	A	X	O	A	O	O	X
Capacity planning			O	V	A	O	X	O	O
Transportation				O	O	O	O	O	V
Internationalization					O	A	A	O	A
Supply chain integration						O	V	A	O
Technological advances							O	V	O
Employee turnover rate								O	O
Demand forecast									O
Information sharing in table above procurement source									

5.4.2 Formulation of reachability matrix

The SSI matrix is changed into a 0, 1 binary form of matrix as presented in Table 5.4, by replacing V, A, X and O symbols with 1 and 0. The conversion is based on the following arrangement:

- If (i, j) value in the SSI matrix is represented by V, then in the reachability matrix, (i, j) is represented by 1 and (j, i) is represented by 0.
- If (i, j) value in the SSI matrix is represented by A, then in the reachability matrix, (i, j) is represented by 0 and (j, i) is represented by 1.
- If (i, j) value in the SSI matrix is represented by X, then in the reachability matrix both (i, j) and (j, i) are represented by 1.
- If (i, j) value in the SSI matrix is represented by O, then in the reachability matrix both (i, j) and (j, i) are represented by 0.

5.4.3 Obtaining final reachability matrix

Table 5.5 presents the details of reachability matrix obtained initially by considering the transitive relationship among dimensions. Table 5.6 shows the final reachability matrix obtained from the initial reachability matrix with driving and dependence powers.

5.4.4 Level partitioning

After creation of the final reachability matrix, further processing is done to make the hierarchical model based on association among the dimensions. To do this, the sets of reachability and antecedent for each dimension are obtained for the identified dimensions and associated levels. The dimensions for which both sets of reachability and sets of intersection are the same are positioned at the top in the ISM hierarchy. The process is repeated

Table 5.4 Initial reachability matrix

Dimensions	1	2	3	4	5	6	7	8	9	10	Driving power
Supplier complexity	1	0	0	1	0	0	0	0	1	0	3
Capacity planning	0	1	0	1	0	0	1	0	0	0	3
Transportation	1	0	1	0	0	0	0	0	1	0	3
Internationalization	1	0	0	1	0	0	0	0	0	0	2
Supply chain integration	0	1	0	0	1	0	1	0	0	0	3
Technological advances	1	0	0	1	0	1	0	1	0	1	5
Employee turnover rate	0	1	0	1	0	0	1	0	0	0	3
Demand forecast	0	0	0	0	1	0	0	1	0	0	2
Procurement source	1	0	0	1	0	0	0	0	1	0	3
Information sharing	0	0	0	0	0	0	1	1	0	1	3
Dependence power	5	3	1	6	2	1	4	3	3	2	

Table 5.5 Initial reachability matrix with transitive relationships

Dimensions	1	4	9	2	3	7	5	8	10	6	Driving power	Level
1	1	1	1	0	0	0	0	0	0	0	3	1
4	1	1	1*	0	0	0	0	0	0	0	3	1
9	1	1	1	0	0	0	0	0	0	0	3	1
2	1*	1	1*	1	0	1	0	0	0	0	5	2
3	1	1*	1	0	1	0	0	0	0	0	4	2
7	1*	1	1*	1	0	1	0	0	0	0	5	2
5	1*	1*	1*	1	0	1	1	0	0	0	6	3
8	1*	1*	1*	1*	0	1*	1	1	0	0	7	4
10	1*	1*	1*	1*	0	1	1*	1	1	0	8	5
6	1	1	1*	1*	0	1*	1*	1	1	1	9	6
Dependence power	10	10	10	6	1	6	4	3	2	1		
Level	1	1	1	2	2	2	3	4	5	6		

until all levels of the structural model are determined. In this case, the level identification is completed using four iterations for the nine dimensions of offshore outsourcing. All these four iterations are as shown in Table 5.7.

5.4.5 Building of the ISM model

Based on the final result of iterations, an initial diagraph is obtained which includes transitivity links. After eliminating the transitivity links, a diagraph is drawn. The ISM model as developed from the digraph representing

Table 5.6 Final reachability matrix

Dimensions	1	2	3	4	5	6	7	8	9	10	Driving power
Supplier complexity	1	0	0	1	0	0	0	0	1	0	3
Capacity planning	1*	1	0	1	0	0	1	0	1*	0	5
Transportation	1	0	1	1*	0	0	0	0	1	0	4
Internationali-zation	1	0	0	1	0	0	0	0	1*	0	3
Supply chain integration	1*	1	0	1*	1	0	1	0	1*	0	6
Technological advances	1	1*	0	1	1*	1	1*	1	1*	1	9
Employee turnover rate	1*	1	0	1	0	0	1	0	1*	0	5
Demand forecast	1*	1*	0	1*	1	0	1*	1	1*	0	7
Procurement source	1	0	0	1	0	0	0	0	1	0	3
Information sharing	1*	1*	0	1*	1*	0	1	1	1*	1	8
Dependence power	10	6	1	10	4	1	6	3	10	2	

Table 5.7 Iterations

		Iteration I		
S. No.	Reachability set	Antecedent set	Intersection set	Level
1	1, 4, 9	1, 2, 3, 4, 5, 6, 7, 8, 9, 10	1, 4, 9	I
2	1, 2, 4, 7, 9	2, 5, 6, 7, 8, 10	2, 7	
3	1, 3, 4, 9	3	3,	
4	1, 4, 9	1, 2, 3, 4, 5, 6, 7, 8, 9, 10	1, 4, 9,	I
5	1, 2, 4, 5, 7, 9	5, 6, 8, 10	5,	
6	1, 2, 4, 5, 6, 7, 8, 9, 10	6	6,	
7	1, 2, 4, 7, 9	2, 5, 6, 7, 8, 10	2, 7	
8	1, 2, 4, 5, 7, 8, 9	6, 8, 10,	8	
9	1, 4, 9	1, 2, 3, 4, 5, 6, 7, 8, 9, 10,	1, 4, 9	I
10	1, 2, 4, 5, 7, 8, 9, 10	6, 10	10	

the hierarchy of dimensions is presented in Figure 5.2. Dimension at Level I is positioned at the top in the model followed by dimensions on the second position at Level II, dimensions on the third position at Level III and similarly the variables are placed up to the last level one by one.

LEVEL

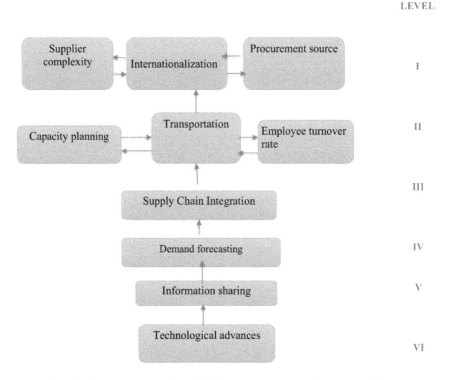

Figure 5.2 Qualitative structural model for perfect order fulfilment in the upstream supply chain.

5.5 MODEL ANALYSIS AND VALIDATION

To assess the power of dimensions, viz. driver and dependent power and to further classify them under different clusters, namely, autonomous, linkage, dependent and independent, MICMAC analysis as discussed in Section 2.1 is used. Figure 5.3 presents the results of MICMAC analysis which is used to validate the ISM model.

The four clusters with definitions are discussed as under:

1. *Autonomous dimensions*: Those dimensions possessing the weakest powers of driving and dependence fall in the group of autonomous dimensions.
2. *Linkage dimensions*: The dimensions having the strongest powers of driving and dependence fall in the group of linkage dimensions.
3. *Dependent risk dimensions*: This category contains criteria which possess weak driving and strong dependence powers.
4. *Independent risk dimensions*: Here in this category, dimensions possess strong power of driving and weak power of dependence.

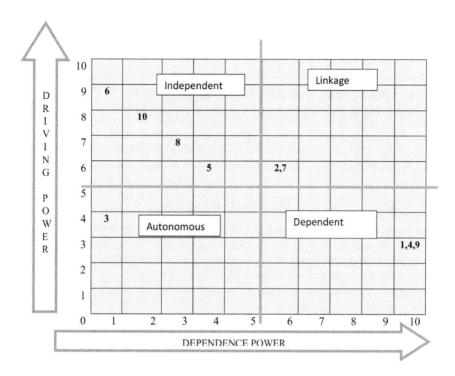

Figure 5.3 MICMAC classification diagram for perfect order fulfilment variables.

The four quadrants in Figure 5.3 present these dimensions and are briefly discussed as under.

Quadrant 1: In the study, two POF dimensions, viz. capacity planning and employee turnover falls into linkage variables.

Quadrant 2: In the study, four POF dimensions, viz. technological advances, information processing, demand forecasting and supply chain integration falls into this cluster. All of these possess high driving power. Use of the latest technology for information processing such as radio frequency identification (RFID), electronic data interchange (EDI), use of material requirement planning (MRP) software helps vendors to manage POF in a much better manner.

Quadrant 3: In the study, transportation is placed under an autonomous category. As this dimension is important and regulates third-party logistics (3PL), fourth-party logistics (4PL) services in an effective manner.

Quadrant 4: In the study, three POF dimensions, viz. supplier complexity, internationalization and procurement source are placed in this cluster. All three dimensions are related with each other. Globalization offers wide procurement opportunities but at the same time supplier section becomes critical owing to different cost, quality and service dimensions.

5.6 HIERARCHAL STRUCTURAL MODEL FOR POF IN DOWNSTREAM SUPPLY CHAIN

The model has been developed by following various steps as shown in Figure 5.1 and discussed in Section 5.3.

5.6.1 Structural self-interaction matrix

The relative association among the dimensions is represented as "lead to", which states that how one dimension leads to another dimension. Consultations with a total of six experts, out of which four were from the outsourcing companies and two were from academic units were made to formulate the contextual relation among the upstream dimensions. On the basis of responses, SSI matrix is established as tabulated in Table 5.8. Symbols such as V, A, X and O are used to make the SSI matrix.

5.6.2 Formation of the initial reach ability matrix

The SSI matrix is converted into 0, 1 matrix called as the initial reachability matrix (IRM). The symbols representing the relationship among

Table 5.8 SSI matrix for dimensions

S. No.	Dimensions	11	10	9	8	7	6	5	4	3	2	1
1	Communication and information sharing	V	A	O	A	V	X	V	V	V	V	X
2	Product quality	O	V	X	A	X	A	A	V	O	X	
3	Demand forecast	X	A	O	A	A	V	X	A	X		
4	Lead time	A	O	V	A	X	A	X	X			
5	Inventory	A	O	V	A	A	A	X				
6	Precise data	O	A	O	A	V	X					
7	Product volume	A	A	V	A	X						
8	Supply chain tiers	V	O	V	X							
9	Responsiveness	A	O	X								
10	Client feedback	O	X									
11	Shipping duration	X										

the dimensions in Table 5.8 are substituted by 0 and 1 by following the rules:

(i) If (i, j) value in the SSI matrix is represented by V, then (i, j) value in RM will be 1 and (j, i) value is 0.
(ii) If (i, j) value in the SSI matrix is represented by A, then (i, j) value in RM will be 0 and (j, i) value is 1.
(iii) If (i, j) value in the SSI matrix is represented by X, then both (i, j) and (j, i) values in RM are 1.
(iv) If (i, j) value in the SSI matrix is represented by O, both (i, j) and (j, i) entries in RM are 0.

The IRM is shown in Table 5.9.

5.6.3 Development of the final reachability matrix

The transitive relation between the variables is eliminated and reachability matrix is obtained finally from the initially obtained matrix. Table 5.10 shows the entries in RM along with the driver and dependence scores of various dimensions. The last column presents the values of driving power and the last row presents the values of dependence power of dimensions.

5.6.4 Level partitioning

After creation of the reachability matrix finally, further processing is done to make the hierarchical model based on association among the criteria. To do this, the sets of reachability and antecedent for each dimension are obtained for the identified dimensions and associated levels. The dimensions for which both sets of reachability and sets of

Table 5.9 IRM matrix

S. No.	Dimensions	11	10	9	8	7	6	5	4	3	2	1
I	Communication	I	0	0	0	I	I	I	I	I	I	I
2	Product quality	0	I	I	0	I	0	0	I	0	I	0
3	Demand forecast	I	0	0	0	0	I	I	0	I	0	0
4	Lead time	0	0	I	0	I	0	I	I	I	0	0
5	Inventory	0	0	I	0	0	0	I	I	I	I	0
6	Precise data	0	0	0	0	I	I	I	I	0	I	I
7	Product volume	0	0	I	0	I	0	I	I	I	I	0
8	Supply chain tiers	I	0	I	I	I	I	I	I	I	I	I
9	Responsiveness	0	0	I	0	0	0	0	0	0	I	0
10	Client feedback	0	I	0	0	I	I	0	0	I	0	I
11	Shipping duration	I	0	I	0	I	0	I	I	I	0	0

Table 5.10 Final reachability matrix

S. No.	Dimensions	1	2	3	4	5	6	7	8	9	10	11
1	Communication	1	1	1	1	1	1	1	0	0	0	1
2	Product quality	0	1	0	1	0	0	1	0	1	1	0
3	Demand forecast	0	0	1	0	1	1	0	0	0	0	1
4	Lead time	0	0	1	1	1	0	1	0	1	0	0
5	Inventory	0	1	1	1	1	0	0	0	1	0	0
6	Precise data	1	1	0	1	1	1	1	0	0	0	0
7	Product volume	0	1	1	1	1	0	1	0	1	0	0
8	Supply chain tiers	1	1	1	1	1	1	1	1	1	0	1
9	Responsiveness	0	1	0	0	0	0	0	0	1	0	0
10	Client feedback	1	0	1	0	0	1	1	0	0	1	0
11	Shipping duration	0	0	1	1	1	0	1	0	1	0	1

intersection are the same are positioned at the topmost position in the ISM model. The procedure is repeated till all levels of the structural model are determined. In this case, the level identification is completed using six iterations for the nine dimensions. The iterative results are presented in Table 5.11.

5.6.5 Formulation of conical matrix

It is formulated by joining together dimensions which are at a similar level in corresponding rows and columns of the final reachability matrix (FRM). Table 5.12 shows the elements in a conical matrix. Thereafter, the total number of ones are aggregated in the rows to determine the power of driving and in the same manner the total count of ones are aggregated in the columns up to obtain the power of dependency. Based upon the sums, ranks of dimensions are obtained.

5.6.6 Development of digraph

The ISM model is made after developing the conical matrix. Initially, digraph is made from the conical matrix by eliminating transitivity links, and finally the digraph showing the relations among dimensions is made as presented in Figure 5.4a.

5.6.7 ISM model development and MICMAC analysis

The ISM model as developed from digraph representing the hierarchy of dimensions is presented in Figure 5.4b. Dimension at Level I is positioned at the top in the model followed by dimensions on the second position at Level II, dimensions on the third position at Level III and

Table 5.11 Iterations and levels

S. No.	Dimensions	Reachability	Antecedent	Intersection	level
Iteration I					
1	Communication	1,2,3,4,5,6,7,11	1,6,8,10	1,6	
2	Product quality	2,4,7,9,10	1,2,5,6,7,8,9	2,7,9	
3	Demand forecast	3,5,6,11	1,3,4,5,7,8,10,11	3,5,11	
4	Lead time	3,4,5,7,9	1,2,4,5,6,7,8,11	4,5,7	
5	Inventory	2,3,4,5,9	1,3,4,5,6,7,8,11	3,4,5	
6	Precise data	1,2,4,5,6,7	1,3,6,8,10	1,6	
7	Product volume	2,3,4,5,7,9	1,2,4,6,7,8,10,11	2,4,7	
8	Supply chain tiers	1,2,3,4,5,6,7,8,9,11	8	8	
9	Responsiveness	2,9	2,4,5,7,8,9,11	2,9	I
10	Client feedback	1,3,6,7,10	2,10	10	
11	Shipping duration	3,4,5,7,9,11	1,3,8,11	3,11	
Iteration II					
1	Communication and information sharing	1,3,4,5,6,7,11	1,6,8,10	1,6	
3	Demand forecast	3,5,6,11	1,3,4,5,7,8,10,11	3,5,11	
4	Lead time	3,4,5,7	1,4,5,6,7,8,11	4,5,7	
5	Inventory	3,4,5	1,3,4,5,6,7,8,11	3,4,5	2
6	Precise data	1,4,5,6,7	1,3,6,8,10	1,6	
7	Product volume	3,4,5,7	1,4,6,7,8,10,11	4,7	
8	Supply chain tiers	1,3,4,5,6,7,8,11	8	8	
10	Client feedback	1,3,6,7,10	10	10	
11	Shipping duration	3,4,5,7,11	1,3,8,11	3,11	
Iteration III					
1	Communication and information sharing	1,4,6,7,11	1,6,8,10	1,6	
4	Lead time	4,7	1,4,6,7,8,11	4,7	3
6	Precise data	1,4,6,7	1,6,8,10	1,6	
7	Product volume	4,7	1,4,6,7,8,10,11	4,7	3
8	Supply chain tiers	1,4,6,7,8,11	8	8	
10	Client feedback	1,6,7,10	10	10	
11	Shipping duration	4,7,11	1,8,11	3,11	
Iteration IV					
1	Communication and information sharing	1,6,11	1,6,8,10	1,6	

(Continued)

Table 5.11 (Continued) Iterations and levels

S. No.	Dimensions	Reachability	Antecedent	Intersection	level
6	Precise data	1,6	1,6,8,10	1,6	4
8	Supply chain tiers	1,6,8,11	8	8	
10	Client feedback	1,6,10	10	10	
11	Shipping duration	11	1,8,11	3,11	4
Iteration V					
1	Communication and information sharing	1	1,8,10	1	5
8	Supply chain tiers	1,8	8	8	
10	Client feedback	1,10	10	10	
Iteration VI					
8	Supply chain tiers	8	8	8	6
10	Client feedback	10	10	10	6

Table 5.12 Dimensions in matrix

S. No.	Dimensions	2	9	3	5	4	7	6	11	1	8	10	Driving power
2	Product quality	1	1	0	0	1	1	0	0	0	0	1	5
9	Responsiveness	1	1	0	0	0	0	0	0	0	0	0	2
3	Demand forecast	0	0	1	1	0	0	1	1	0	0	0	4
5	Inventory	1	1	1	1	1	0	0	0	0	0	0	5
4	Lead time	0	1	1	1	1	1	0	0	0	0	0	5
7	Product volume	1	1	1	1	1	1	0	0	0	0	0	6
6	Precise data	1	0	0	1	1	1	1	0	1	0	0	6
11	Shipping duration	0	1	1	1	1	1	0	1	0	0	0	6
1	Communication	1	0	1	1	1	1	1	1	1	0	0	8
8	Supply chain tiers	1	1	1	1	1	1	1	1	1	1	0	10
10	Client feedback	0	0	1	0	0	1	1	0	1	0	1	5
	Dependence power	7	7	8	8	8	8	5	4	4	1	2	62/62

similarly the variables are placed up to the last level one by one. In the study, two dimensions are placed at Level I to Level III and one each from Level IV to Level VI. The model shows the dimensions and their positions at various levels and thus depicts the inter-association between the dimensions. It indicates that the internationalization and technological advances possess high driving power and less dependence power.

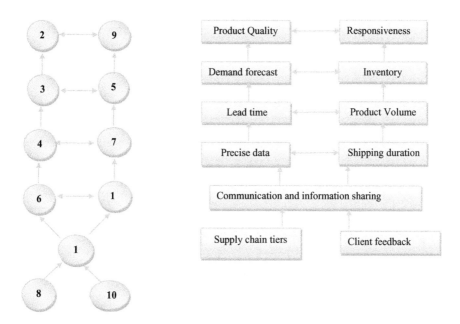

Figure 5.4 (a) Diagraph and (b) interpretive structure model.

5.7 MODEL ANALYSIS AND VALIDATION

To assess the driver and dependent power and to further classify the dimensions under different clusters, viz. dependent, autonomous, independent and linkage, MICMAC diagram as discussed in Section 2.1 is used. MICMAC diagram as discussed in Section 2.1 is used for analysis. Figure 5.5 presents the results of MICMAC analysis which is used to validate the ISM model for vendor selection dimensions.

The four clusters with definitions are discussed as under:

1. *Autonomous dimensions:* Those dimensions possessing the weakest powers of driving and dependence fall in the group of autonomous dimensions.
2. *Linkage dimensions*: The dimensions having the strongest powers of driving and dependence falls in the group of linkage dimensions.
3. *Dependent risk dimensions*: This category contains criteria which have weak driving and strong dependence powers.
4. *Independent risk dimensions*: Here, in this category, dimensions possess strong power of driving and weak power of dependence.

The four quadrants in Figure 5.5 present these four groups of dimensions and are briefly discussed as under:

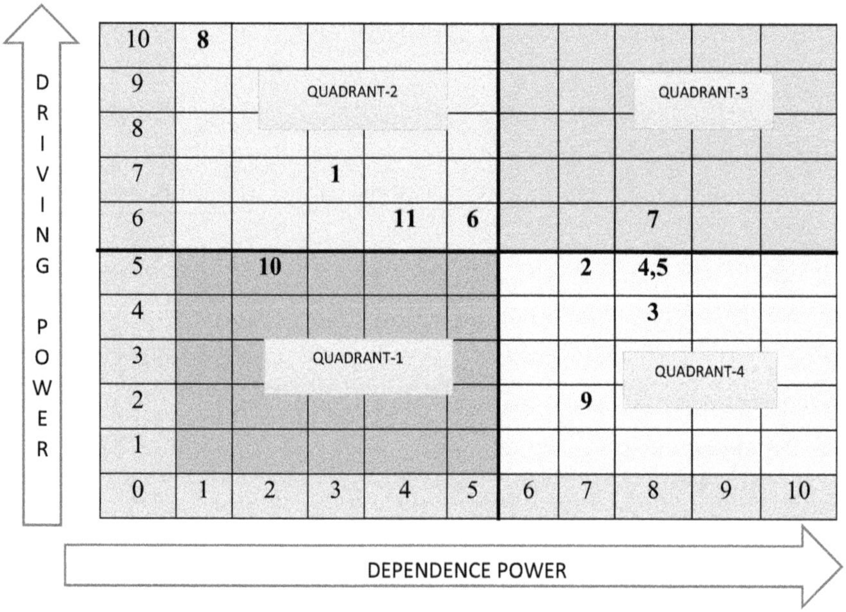

MICMAC diagram (DRIVING POWER on vertical axis, DEPENDENCE POWER on horizontal axis):

DRIVING POWER \ DEPENDENCE POWER	1	2	3	4	5	6	7	8	9	10
10	8									
9		QUADRANT-2						QUADRANT-3		
8										
7		1								
6				11	6			7		
5		10					2	4,5		
4								3		
3		QUADRANT-1						QUADRANT-4		
2							9			
1										
0										

Figure 5.5 MICMAC diagram.

- *Quadrant 1*: POF dimension client feedback/feedback comes under this quadrant.
- *Quadrant 2*: Four POF dimensions, viz. product quality, demand forecast, lead time, inventory and responsiveness fall into this quadrant.
- *Quadrant 3*: POF dimension product volume comes under linkage.
- *Quadrant 4*: Four POF dimensions, viz. communication and information sharing, precise data, supply chain tiers and shipping duration, fall under this quadrant.

5.8 CONCLUSION

In the present times, e-commerce companies enjoy a competitive advantage over traditional companies because of their ability to please the customer with low price tags, wider choices and smoother delivery schedules. This is because of various POF strategies being adopted by them such as distributed delivery centre–based fulfilment, build to order, third-party (3P) order fulfilment, drop shipping, etc. Amazon is a perfect example of e-commerce retailer following the drop shipping strategy. On the other hand, the traditional methods for order fulfilment are distribution centre–based order execution, warehousing and vendor-based direct fulfilment. To increase supply chain efficiency, POF is an important part of supply chain logistics. It helps to measure the performance of supply chain in shipping the right product, at the right place, at the right time, in the right condition and right packaging, with the right quantity to the end customer.

To this effect, the contributions of the present chapter are: (i) determination of the POF dimensions for both upstream and downstream supply chain logistics, (ii) modeling the contextual association between the dimensions and development of the structural model and (iii) exploration of influential dynamics, i.e., driver-dependence degree between the variables. The study provides a thorough understanding of various dimensions that impact POF in upstream supply chain. The developed ISM model for upstream supply chain would help the managers to recognize the significance of these dimensions, i.e., to assist them to pinpoint the key dimensions for better supply chain performance.

A major finding of this chapter lies in the determination of hierarchal levels among key dimensions that affect the POF. The dimensions, viz. supply chain integration and transportation have higher driving power and thus considerably impact POF in an upstream supply chain. These dimensions are placed in the lowest level of hierarchy and are supposed to be main cause for other dimensions. Today businesses are going global, so companies cannot work in isolation or neglect external factors based on economic trends, competitive situations or technology innovation in other countries. Therefore, in recent times, with the advent of internationalization, the demand for partnerships and outsourcing has risen, which calls for developing the supply chain not only through own resources but also with multiple suppliers for POF. Thus, top management should consider these dimensions in their strategic plans.

Working with the same approach, the second part of the chapter identifies the dimensions responsible for fulfilling the orders in a downstream SC. The dimensions are analysed and a model is built by applying the ISM method and to analyse the dynamics of the dimensions of the MICMAC diagram is used. From this MICMAC diagram, it is found that the dimensions precise data, shipping duration, number of echelons and communication and information sharing possess the highest power of driving. Four dimensions, viz. product quality, demand forecast, lead time, inventory and responsiveness possess high dependence power, hence firms should optimize the resources to make their supply chains more responsive and efficient.

The developed models are not restricted to specific industrial sectors or organizations. The models aim to provide a deeper understanding of the dimensions which may help the supply chain managers or supply chain practitioners to undertake strategic, tactical and operational decisions.

FURTHER READING

Akhtar, M., Panda, P.B., Khan, A.A. (2021). An integrated fuzzy multi-criteria group decision-making approach for supplier evaluation and selection in oil and gas industry. *International Journal of Business Performance and Supply Chain Modelling*, 12(2), 147–178.

Bhattacharya, A., Chetty, P. (2019). Order fulfilment strategies in supply chain management. [online] Project Guru. Available at: https://www.projectguru.in/order-fulfilment-strategies-supply-chain-management/ [Accessed 03 Dec. 2022].

Lakri, S., Dallery, Y., Jemai, Z. (2015). Measurement and management of supply chain performance: practices in today's large companies. *Supply Chain Forum: An International Journal*, 16, 16–30.

Najmi, A., Gholamian, M.R., Makui, A. (2013). Supply chain performance models: a literature review on approaches, techniques, and criteria. *Journal of Operations and Supply Chain Management*, 6, 94–113.

Puska, A., Kozarevic, S., Okicic, J. (2020). Investigating and analyzing the supply chain practices and performance in agro-food industry. *International Journal of Management Science and Engineering Management*, 15(1), 9–16. doi:10.1080/17509653.2019.1582367.

Sharma, R.K. (2021). ISM and fuzzy logic approach to model and analyze the variables in downstream supply chain for perfect order fulfillment. *International Journal of Quality & Reliability Management*, 38(8), 1722–1746. https://doi.org/10.1108/IJQRM-09-2020-0294.

Sharma, R.K. (2023). Assessment of interaction among key variables responsible for perfect order fulfilment in an upstream supply chain network. *International Journal of Business Performance and Supply Chain Modelling*, 13, 380–404. https://doi.org/10.1504/IJBPSCM.2022.128126.

Sheel, A., Singh, Y.P., Nath, V. (2020). Managing agility in the downstream petroleum supply chain. *International Journal of Business Excellence*, 20(2), 269–294.

Zekhnini, K., Cherrafi, A., Bouhaddou, I., Benghabrit, Y., Garza-Reyes, J.A. (2019). Supply chain management 4.0: literature review and research framework. *Benchmarking: An International Journal*, 28, 1–52.

Chapter 6

Modeling dimensions associated with location of warehouse facility in logistics

6.1 INTRODUCTION

Worldwide, companies accrue an expenditure of about $350 billion a year on warehousing operations and this number is growing each year as pick sizes shrink and costs become high, which raises pressure not only on margins but also on meeting the service levels. With advancements in technology, warehousing operations are becoming simpler. The growth of ecommerce has led to a proliferation of stock keeping units, and there's an ever-growing need to please the customers by providing order fulfilment. The design of supply chain network and its management is incomplete without determining the location, design and management of warehouses. In present times, warehouses operate not only as centres for storage but also as centres for value-addition activities, viz. assembly, packaging and repair operations within their premises. In the supply chain, goods are to be transported to the end customer in the right condition and at the right time with the right price. To meet these challenges, it is important for the logistics firms to build an effective locational distribution infrastructure.

The infrastructure consists of freight transport and storage system which help in movement of goods from the manufacturing point to consumption point. These distribution structures are classified as centralized and decentralized centres. Centralized structures comprise single distribution centre location with direct shipping to the customers, while decentralized distribution structures include multiple distribution locations. For example, Dell a personal computer (PC) manufacturer uses direct shipping for transporting goods to their private clients, while Zara follows a decentralized distribution system by addition of a new distribution centre (DC) in the Netherlands. Also, Amazon with its 1300 local distribution centres near European cities possess a decentralized distribution structure (Chopra 2003; Ecommerce News 2017). Thus, distribution logistics' primary goal is to create equilibrium between the supplies of goods and customer demand.

DOI: 10.1201/9781032707884-6

Existing literature shows that understanding warehouse design and management principles plays a crucial role in increasing the efficiency of warehouse operations, reduction in worker fatigue and improvement in service level. It is important for the supply chain managers to have knowledge of the factors which govern the selection of warehouse decisions for their companies. This helps them in various ways. (i) It helps to meet service levels as desired by the end customer, (ii) it lowers the cost associated with logistics and (iii) it offers agility in meeting customer demands. There are various trade-off decisions in distribution logistics, such as between service level and logistical costs. Also, high inventory costs make firms to go for centralized distribution because it minimizes the number of storage locations. At the same time, higher transportation cost makes the firms to choose decentralized distribution as it minimizes transportation cost.

6.2 WAREHOUSE FACILITY LOCATION DECISION DIMENSIONS IN LOGISTICS

Logistics provides a fundamental mechanism to manage the complexity of global operations between key stakeholders, viz. buyers, suppliers and customers (Mentzer et al. 2001). The logistics expenditures, which include mainly transportation and warehousing activities, represent about 12% of global shipper sales revenues (Langley and Capgemini 2012). As companies focus on their core competencies and outsource their non-core activities to suppliers (Christopher 2011), logistics is one of the key functions being outsourced. The main dimensions considered for locating warehousing activities in the chapter are based on studies undertaken by Kumar et al. (2021), Onstein et al. (2019), Heitz et al. (2019), Christopher (2011), Chopra (2003) and Higginson and Bookbinder (2005). These dimensions are related to logistical costs, demand and delivery readiness, product characteristics, human resource, location approachability, competitive strategy and related factors. The various dimensions along with the sub-dimensions and description are presented in Table 6.1.

6.3 CONCEPTUAL FRAMEWORK FOR WAREHOUSE LOCATION SELECTION DIMENSIONS IN LOGISTICS

The present section presents details of conceptual framework and methodology used. Figure 6.1 presents the framework. The proposed framework consists of two phases:

Table 6.1 Warehouse location decision dimensions in logistics

S. No.	Dimensions	Sub-dimensions	Description
1	Demand level	A1 Product demand and pattern	Demand level depends upon temporal and spatial patterns. For high volume and spatially dispersed product demands, multiple location of distribution centres is preferred over the centralized structure.
2	Delivery readiness	B1 Vendor lead time B2 Response time B3 Vendor reliability B4 Returns and repairs	In location decisions, the main dimension is lead time or delivery readiness. For decentralized structure with multiple locations of distribution centres, the delivery time is shortened with increased cost of logistics. It also improves return and repairs.
3	Product characteristics	C1 Product value C2 Product quantity C3 Perishable–non-perishable	The product characteristics, viz. product value, its volume and nature (perishable–non-perishable) affect the location decision.
4	Logistical costs	D1 Inbound and outbound costs D2 Process logistical costs D3 Inventory cost D4 Warehouse automation	Logistics costs which affect the location decisions are costs associated with inbound and outbound logistics. Also, the costs are inventory-related costs, viz. ordering and carrying costs. This involves handling and storage costs with warehousing.
5	Human resource	E1 Availability of labour E2 Training and Education	Location decisions are also dependent upon availability of human labour. Training and education of human resource also plays an important role in location specific decisions.
6	Location approachability	F1 Terrain F2 Distance	The type of terrain and distance of location of distribution centres. Location approachability also depends upon the position of various stakeholders in the supply chain.
7	Competitive strategy	G1 Customer relationship G2 Operational superiority G3 Product leadership	The competitive strategy being adopted by businesses their relationship with customers and product leadership with operational superiority also effects the locational logistics.
8	Related factors	H1 Zone regulations H2 Custom duties H3 Local taxes and govt. subsidies H4 Foreign trade regulations H5 Environmental conditions H6 Insurance policy	The related factors which influence the location are local taxes, foreign trade regulations and custom duties. These factors are strongly associated with accessibility, labour and land availability, and hence influence costs associated with logistics in a direct manner.

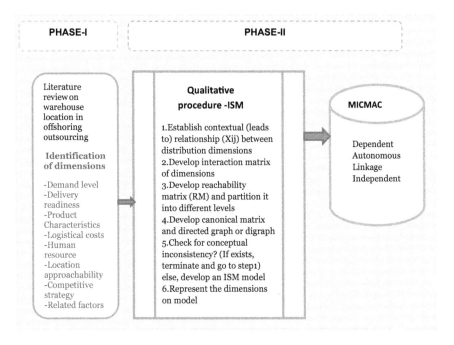

Figure 6.1 Theoretical outline for warehouse location selection decision.

- In phase I, various dimensions associated with influencing distribution structure decisions have been identified through the literature and conversation with domain experts.
- In phase II, an interpretive structural modeling (ISM) approach is used to model the complex relationship among location centre selection dimensions. Also the Matrice d'Impacts Croisés Multiplication Applied to a Classification (MICMAC) diagram is constructed to examine the driver and dependence power of dimensions. This is accomplished by portraying these dimensions under four groups, i.e., dependent, independent, linkage and autonomous.

6.4 HIERARCHAL STRUCTURAL MODEL FOR DIMENSIONS ASSOCIATED WITH WAREHOUSE LOCATION SELECTION

6.4.1 Structural self-interaction matrix (SSIM)

The relative association among the dimensions is represented as "lead to", which states how one dimension leads to another dimension. Consultations with a total of six experts, out of which four were from the companies and

two were from academic units were made to formulate the contextual relation among the upstream dimensions. On the basis of responses, structured
self-interaction (SSI) matrix is made as presented in Table 6.2. The matrix
was formed using V, A, X, and O symbols.

6.4.2 Reachability matrix

The SSI matrix is converted into (0, 1) matrix, known as initial reachability
matrix (IRM) as shown in Table 6.3. The symbols representing the relationship among the dimensions in Table 6.2 are substituted by 0 and 1 by
following the rules:

    i. If (i, j) value in SSI matrix is represented by V, then (i, j) value in RM
       will be 1 and (j, i) value is 0.
    ii. If (i, j) value in the SSI matrix is represented by A, then (i, j) value in
       RM will be 0 and (j, i) value is 1.

Table 6.2 Self-interaction matrix for dimensions

S. No.	Dimensions	8	7	6	5	4	3	2	1
1	Demand level	O	O	O	O	X	O	X	–
2	Delivery readiness	A	A	A	O	X	A	–	
3	Product characteristics	O	V	O	O	V	–		
4	Logistical costs	A	A	A	A	–			
5	Human resource	A	V	A	–				
6	Location approachability	A	V	–					
7	Competitive strategy	A	–						
8	Related factors	–							

Table 6.3 Reachability matrix

Dimensions	1	2	3	4	5	6	7	8
1	1	1	0	1	0	0	0	0
2	1	1	0	1	0	0	0	0
3	0	1	1	1	0	0	1	0
4	1	1	0	1	0	0	0	0
5	0	0	0	1	1	0	1	0
6	0	1	0	1	1	1	1	0
7	0	1	0	1	0	0	1	0
8	0	1	0	1	1	1	1	1

iii. If (i, j) value in the SSI matrix is represented by X, then both (i, j) and (j, i) values in RM will be 1.

iv. If (i, j) value in the SSI matrix is represented by O, both (i, j) and (j, i) entries in RM are 0.

6.4.3 Development of final reachability matrix

The transitive relation between the dimensions is eliminated and reachability matrix is obtained in the final from the initial one. Table 6.4 shows the entries in the matrix with the driver and dependence scores of various dimensions. The last column presents the values of driving power and last row presents the values of dependence power of variables. For the problem, the transitivity links determined are 5-4-1, 5-4-2, etc. In Table 6.4, to show the transitivity, 1* is used for the dimensions which follow it. The final reachability matrix so obtained after transitivity check is shown in Table 6.4.

6.4.4 Level partition

After creation of the final reachability matrix, further processing is done to make the hierarchical model based on association among the criteria. To do this, the sets of reachability and antecedent for each dimension are obtained for the identified dimensions and associated levels. The dimensions for which both sets of reachability and sets of intersection are the same are positioned at the top in the ISM hierarchy. The procedure is repeated until all levels of the structural model are determined. In this case, the level identification is completed using eight iterations for the nine dimensions. Table 6.5 shows the iterative steps.

6.4.5 Formulation of conical matrix

The conical matrix is formulated by joining together risk dimensions which are at a similar level in corresponding rows and columns of the final

Table 6.4 Reachability matrix with transitive links

Dimensions	1	2	3	4	5	6	7	8	Drive power
1	1	1	0	1	0	0	0	0	3
2	1	1	0	1	0	0	0	0	3
3	1*	1	1	1	0	0	1	0	5
4	1	1	0	1	0	0	0	0	3
5	1*	1*	0	1	1	0	1	0	5
6	1*	1	0	1	1	1	1	0	6
7	1*	1	0	1	0	0	1	0	4
8	1*	1	0	1	1	1	1	1	7
Dependent power	8	8	1	8	3	2	5	1	

Table 6.5 Iterative solutions

Dimensions	Reachability	Antecedent	Intersection	Level
First iterative solution				
1	1,2,4	1,2,3,4,5,6,7,8	1,2,4	I
2	1,2,4	1,2,3,4,5,6,7,8	1,2,4	I
3	1,2,3,4,7	3	3	
4	1,2,4	1,2,3,4,5,6,7,8	1,2,4	I
5	1,2,4,5,7	5,6,8	5	
6	1,2,4,5,6,7	6,8	6	
7	1,2,4,7	3,5,6,7,8	7	
8	1,2,4,5,6,7,8	8	8	
Second iterative solution				
3	3,7	3	3	
5	5,7	5,6,8	5	
6	5,6,7	6,8	6	
7	7	3,5,6,7,8	7	II
8	5,6,7,8	8	8	
Third iterative solution				
3	3	3	3	III
5	5	5,6,8	5	III
6	5,6	6,8	6	
8	5,6,8	8	8	
Fourth iterative solution				
6	6	6	6	IV
8	6,8	6,8	8	
Fifth iterative solution				
8	8	8	8	V

reachability matrix (FRM). Table 6.6 shows the elements in the conical matrix. Thereafter, the total number of ones are aggregated in the rows to determine the power of driving and in the same manner the total number of ones are aggregated in the columns up to obtain the power of dependency. Based upon the sums, ranks of dimensions are obtained.

6.4.6 ISM model development

The ISM model as developed from the digraph representing the hierarchy of dimensions is presented in Figure 6.2. Three dimensions related to service, demand level and logistic cost at Level I are placed at the topmost position in the model followed by the dimension competitive strategy on second position at Level II, dimensions on third position at Level III and similarly the variables are placed up to the last level one by one. In the

Table 6.6 Conical matrix

Dimensions	1	2	4	7	3	5	6	8
1	1	1	1	0	0	0	0	0
2	1	1	1	0	0	0	0	0
4	1	1	1	0	0	0	0	0
7	1	1	1	1	0	0	0	0
3	1	1	1	1	1	0	0	0
5	1	1	1	1	0	1	0	0
6	1	1	1	1	0	1	1	0
8	1	1	1	1	0	1	1	1

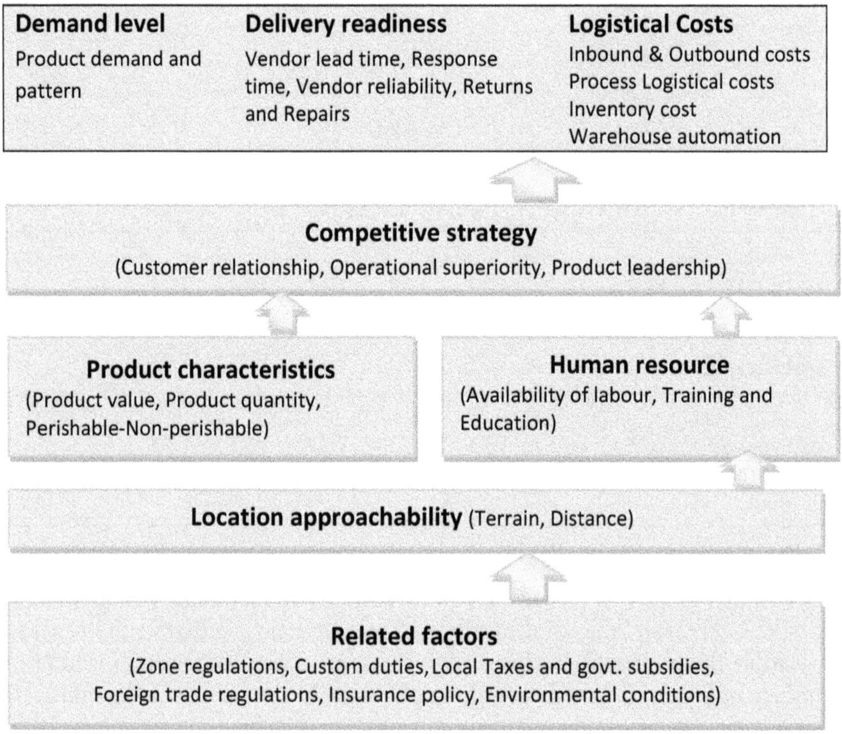

Demand level	**Delivery readiness**	**Logistical Costs**
Product demand and pattern	Vendor lead time, Response time, Vendor reliability, Returns and Repairs	Inbound & Outbound costs Process Logistical costs Inventory cost Warehouse automation

Competitive strategy
(Customer relationship, Operational superiority, Product leadership)

Product characteristics
(Product value, Product quantity, Perishable-Non-perishable)

Human resource
(Availability of labour, Training and Education)

Location approachability (Terrain, Distance)

Related factors
(Zone regulations, Custom duties, Local Taxes and govt. subsidies, Foreign trade regulations, Insurance policy, Environmental conditions)

Figure 6.2 ISM model.

study, two dimensions, product characteristics and human resource, are placed at Level III and location dimension figures at Level IV and related factors dimension at Level V and thus the inter-association between the dimensions are depicted. Competitive strategy depends on three dimensions placed lower to it in the ISM model and it drives the three dimensions (demand level, delivery readiness and logistical costs) above it. Dimension location acts as a driver for all the dimensions above it. The related factors occur at the lowermost level in the ISM model. They drive all other dimensions placed in the model.

6.5 MODEL ANALYSIS AND VALIDATION

To assess the power of dimensions, viz. driver and dependent power and to further classify them under different clusters, namely, autonomous, linkage, dependent and independent, MICMAC analysis as discussed in Section 2.1 is used. Figure 6.3 presents the results of MICMAC analysis which are used to validate the ISM model for vendor selection dimensions.

The four clusters with definitions are discussed as under:

1. *Autonomous dimensions*: Those dimensions possessing the weakest powers of driving and dependence fall in the group of autonomous dimensions.
2. *Linkage dimensions*: The dimensions having the strongest powers of driving and dependence falls in the group of linkage dimensions.
3. *Dependent risk dimensions*: This category contains criteria which possess weak driving and strong dependence powers.
4. *Independent risk dimensions*: Here, in this category, dimensions possess strong power of driving and weak power of dependence.

The four quadrants in Figure 6.3 present these four groups of dimensions and are briefly discussed as under.

- *Quadrant 1*: In the study, no dimension is placed under the autonomous category. Thus, all the dimensions for location dimensions are important.

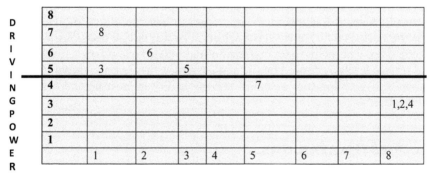

Figure 6.3 MICMAC diagram of dimensions and central vertical line in figure before point 5.

- *Quadrant 2*: In the study, dimensions demand level, delivery readiness, logistical costs and competitive strategy falls into this quadrant.
- *Quadrant 3*: In the study, under this quadrant, no dimension is found.
- *Quadrant 4*: In the study, dimensions like product characteristics, human resource, location approachability and related factors are placed in this quadrant.

6.6 CONCLUSION

Today, in addition to centres for storage, warehouses also perform value-addition activities, viz. assembly, packaging and repair operations within their buildings. In logistics, goods are to be transported to the end customer in the right condition and at the right time with the right price. To meet these challenges, it is important for the logistics firms to build effective locational distribution infrastructure.

The chapter provides description to various dimensions related to warehouse location decisions in supply chain and logistics. The identified dimensions are grouped into eight categories, viz. (i) demand level, (ii) delivery readiness, (ii) product characteristics, (iv) logistical costs, (v) human resource, (vi) location approachability, (vii) competitive strategy and (viii) related factors. The main question answered in this chapter is: *What is the association among the dimensions related to warehouse location decisions in logistics?* To answer this question, a theoretical framework has been developed using the ISM approach which helps to analyse the structural association among the dimensions affecting the decisions. In the ISM model dimensions demand level, delivery readiness, logistical costs, and competitive strategy are positioned at the top level. The dimensions product characteristics and human resource are positioned at the middle level. The dimensions location and related factors are placed at the bottom most position in the ISM model. Results of MICMAC analysis depict the dynamics of these dimensions with respect to their powers of driving and dependence. It helps the supply chain managers to have pre-hand knowledge of the dimensions which govern the selection of warehouse decisions for their companies. This helps them in decision-making in various ways. (i) Meeting up service levels as desired by the end customer, (ii) lowering down the cost associated with logistics and (iii) offering agility in meeting end customer demands.

SUGGESTED READINGS

Chopra, S. (2003). Designing the distribution network in a supply chain. *Transportation Research Part E: Logistics and Transportation Review*, 39(2), 123–140.

De Koster, R.B.M., Johnson, A.L., Roy, D. (2017). Warehouse design and management. *International Journal of Production Research*, 55(21), 6327–6330. https://doi.org/10.1080/00207543.2017.1371856.

Fazlollahtabar, H., Kazemitash, N. (2021). Green supplier selection based on the information system performance evaluation using the integrated best-worst method. *Facta Universitatis*, Series: Mechanical Engineering, 19, 345–360.

Heitz, A., Launay, P., Beziat, A. (2019). Heterogeneity of logistics facilities: an issue for a better understanding and planning of the location of logistics facilities. *European Transport Research Review*, 11(1), 1–20.

Higginson, J.K., Bookbinder, J.H. (2005). Distribution Centres in Supply Chain Operations. In: *Logistics Systems: Design and Optimization* (pp. 67–91). Springer, Boston, MA.

Onstein, A.T., Tavasszy, L.A., van Damme, D.A., (2019). Factors determining distribution structure decisions in logistics: a literature review and research agenda. *Transport Reviews*, 39(2), 243–260.

Mangiaracina, R., Song, G., Perego, A. (2015). Distribution network design: a literature review and a research agenda. *International Journal of Physical Distribution & Logistics Management*, 45(5), 506–531.

Song, G., Sun, L. (2017). Evaluation of factors affecting strategic supply chain network design. *International Journal of Logistics Research and Applications*, 20(5), pp.405–425

Kumar, S., Narkhede, B.E., Jain, K. (2021). Revisiting the warehouse research through an evolutionary lens: a review from 1990 to 2019. *International Journal of Production Research*, 59(11), 3470–3492.

Chapter 7

Modeling human dimensions for mitigating coordination risk in supply chain and logistics

7.1 INTRODUCTION

The impact of various natural and man-made disturbances can be observed over the past few decades on the supply chains (SCs). Also, indirect events, viz. financial crisis, pandemic, etc., too affect the supply chain and logistics performance (Mahdiraji et al. 2022). In industry, dimensions related to human resource impact the execution of supply chain processes significantly. Coordination permits SC partners not only to work together but also to schedule and list their priorities which help to save money and resources too. More recently during COVID 19 pandemic, the contracts among the SC members helped to overcome the delays in meeting the demand for vaccine. Some of the human dimensions cited in literature which may affect the supply chain performance are: management commitment, staff participation, customer relations, corporate social responsibility, organization culture, teamwork, motivation and innovation (Kumar et al. 2019).

Focus by managers on these dimensions help them to coordinate among supply chain entities in an effective manner. The dimensions help to build strong relationships with suppliers and various stakeholders in the supply chain. Supply chain members do not work in isolation as they are dependent on each other to ensure the flow of goods and information. Global offshore outsourcing and advancements in the application of informational and communication technology means in supply chains there is increased dependence among the members. Owing to increased dependence, the uncertainty and risk has increased manifold which may lead to supply chain disruptions. These disruptions result in deviations in supply chain performance and calls for coordination among the members of the supply chain.

DOI: 10.1201/9781032707884-7

7.2 HUMAN DIMENSIONS RESULTING IN LACK OF COORDINATION

Today, supply chains of firms are aggressively competing against each other which makes the role of the human dimension critical to mitigate the risk due to lack of coordination among SC members. For supply chain coordination, ten supply chain human dimensions, viz. management commitment, information sharing (information technology (IT) integration), leadership, corporate and social responsibility, training and learning, communication, customer relationship management, organization culture, collective learning and trust building are considered in the study. All these ten dimensions along with their definitions are presented in Table 7.1

7.3 CONCEPTUAL FRAMEWORK FOR MITIGATING RISK DUE TO LACK OF COORDINATION

In the SC and logistics industry, dimensions related to human resource impact the execution of supply chain processes significantly. Figure 7.1 presents some of the main elements of supply chain coordination. The framework developed for mitigating the risk due to lack of coordination in supply chain consists of two phases as shown in Figure 7.2.

Identification phase: In this phase, there are three steps, viz. supply chain network, supply chain process entities and supply chain human dimensions. Figure 7.2 shows the details of various entities involved in a supply chain process, through upstream and downstream links. The type of flows, viz. material, information and money in a supply chain network is also shown.

ISM approach phase: This phase describes the use of the interpretive structural modeling (ISM) approach for SC coordination. To establish the relative association among human dimensions, the ISM approach is used. In the ISM process, first the dimensions relevant to the problem or issues are identified, which is done through literature study. After identification of dimensions, a context-based association among them is established and written in the matrix form with V, A, X and O as symbols. This symbolic information is used to make a self-interaction matrix with 0–1 as entries representing pair-wise association among dimensions. Entries made in the structural self-interaction matrix (SSI) matrix are changed into a reachability matrix, and the matrix is examined for transitive relationships. Furthermore, the reachability matrix is partitioned into various levels. On the basis of the association of dimensions in the reachability matrix, a digraph is obtained by removing the transitivity. Furthermore, this digraph is changed into an interpretive model. The driving and dependence powers among the dimensions are examined using Matrice d'Impacts Croisés Multiplication Applied to a Classification (MICMAC) analysis.

Table 7.1 Human elements for SC coordination

S. No.	Dimensions	Definitions
1	Management commitment	Management commitment is a strategic dimension which is defined as an obligation of the management to provide resources for coordination among various entities which help the firms to remain competitive in the market.
2	Information sharing (IT integration)	Information sharing deals with dissemination of vital information among supply chain members which helps in coordinating the flow of goods to the end customer with the help of IT integration.
3	Leadership style	Leadership as a human dimension motivates the people to work and focus towards a common goal. It encourages the people to take up new responsibilities and helps in professional development.
4	Corporate and social responsibility	To promote social responsibility of business firms incorporate economic, social and environmental imperatives towards various stakeholders in their supply chain operations.
5	Training	It is a planned and systematic procedure being adopted by firms to grow employee's knowledge and skills through continuous learning and educating them.
6	Communication	Communication within and outside the organization aims for coordination of activities among supply chain members.
7	Customer relationship management	Strategy used by firms to improve and manage relationships with potential customers. It also improves the sales and profit margins and makes supply chains efficient.
8	Organization culture	Organization culture determines the way employees think, react and perform and acts as a key dimension for supply chain coordination.
9	Collective learning	Collective learning is a process in which employees participate together to improve the business operations. It plays the most powerful role in improving supplier and end-customer coordination.
10	Trust building	These actions are undertaken by the management to recognize and mend existing procedures within an organization to build the trust among SC entities within or outside the organization.

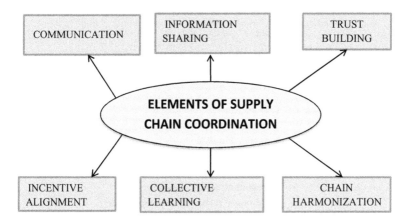

Figure 7.1 Elements of supply chain coordination.

7.4 HIERARCHAL STRUCTURAL MODEL FOR MITIGATING RISK DUE TO LACK OF COORDINATION

This ISM approach to find the structural relationship among various SC human criteria prioritized using the 6σ approach for effective supply chain coordination is discussed in this section. The various steps involved are discussed in the following paragraphs.

7.4.1 Self-interaction matrix for structural relationship

To build a self-interaction matrix, both industry experts and academic experts were considered to find out the contextual relationship between the dimensions. For analyzing a contextual relationship between the dimensions "leads to" the criteria chosen here. The relationship between human dimensions for SC coordination is obtained using four symbols *V, A, X* and *O*.

- *V*=Dimension *i* relates to dimension *j*.
- *A*=Dimension *j* relates to dimension *i*.
- *X*=Dimension *i* and *j* both are related to each other.
- *O*=Dimension *i* and *j* are not related to each other.

The self-interaction matrix showing the relations among human dimensions is developed as shown in Table 7.2. Symbols *V, A, X, O* are used in the SSI matrix by using the logic as discussed: Dimension 1 (management commitment) is related to dimension 10 (process improvement). Thus, the

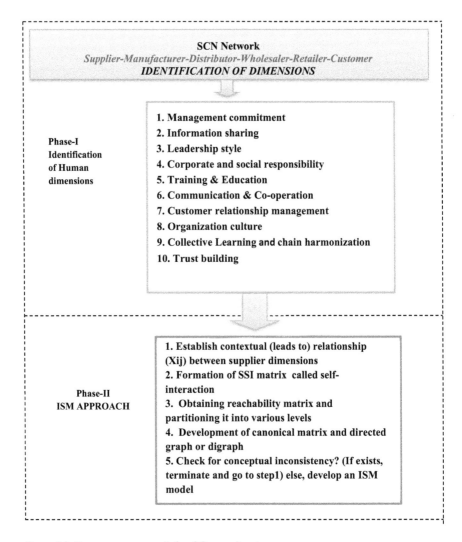

Figure 7.2 Two-stage approach for SC coordination.

association among dimensions 1 and 10 is represented by "*V*" in the SSI matrix, which means dimension 1 leads to dimension 10.

- Dimension 5 (information sharing) can be achieved by dimension 6 (leadership), Thus, the association among the two, viz. dimensions 5 and 6 is shown by "*A*" in the SSI matrix which states that dimension 6 leads to dimension 5.
- Dimension 2 (communication and co-operation) and dimension 5 (information sharing) assist each other. Thus, the association among

Table 7.2 SSI matrix of human dimensions

S. No.	Human dimensions	1	2	3	4	5	6	7	8	9	10
1	Management commitment	V	V	V	V	V	V	V	V	V	V
2	Communication		V	V	X	A	O	V	A	X	
3	Training and Education			V	V	X	V	V	X	V	
4	Customer relationship management				O	V	O	O	A	V	
5	Information sharing and IT integration					A	X	V	O	V	
6	Leadership style						V	V	X	V	
7	Organization culture							X	O	V	
8	Collective learning and chain harmonization								O	V	
9	Corporate and social responsibility									V	
10	Trust building										

the two, i.e., dimensions 2 and 5 is represented using "X" in the SSI, which means dimensions 2 and 5 lead to each other.

- Dimension 4 (customer relationship management) and dimension 7 (team flexibility) are unrelated. Thus, the association among dimensions 4 and 7 is shown by "O" in the SSI matrix, which means dimensions 4 and 7 are not related to each other.

7.4.2 Initial reachability matrix

To formulate the initial reachability matrix, entries represented by V, A, X and O in the self-interaction matrix are replaced by 1 and 0 as per case by following these rules:

i. If (i, j) value in the SSI matrix is represented by V, then (i, j) value in the reachability matrix (RM) will be 1 and (j, i) value is 0.
ii. If (i, j) value in the SSI matrix is represented by A, then (i, j) value in the RM will be 0 and (j, i) value is 1.
iii. If (i, j) value in the SSI matrix is represented by X, then both (i, j) and (j, i) values in the RM is 1.
iv. If (i, j) value in the SSI matrix is represented by O, both (i, j) and (j, i) entries in the RM is 0.

Table 7.3 shows the initial reachability matrix for the considered human dimensions

7.4.3 Final reachability matrix

The transitive relation between the dimensions is eliminated and the reachability matrix is obtained from the initial reachability matrix. Table 7.4

Table 7.3 Initial reachability matrix

S. No.	Human dimensions	1	2	3	4	5	6	7	8	9	10
1	Management commitment	1	1	1	1	1	1	1	1	1	1
2	Communication	0	1	1	1	1	1	1	1	1	1
3	Training and education	0	0	1	1	1	1	1	1	1	1
4	Customer relationship management	0	0	0	1	0	1	0	0	0	1
5	Information sharing	0	0	0	0	1	0	1	1	0	1
6	Leadership	0	0	0	0	0	1	1	1	1	1
7	Organization culture	0	0	0	0	0	0	1	1	0	1
8	Collective learning and chain harmonization	0	0	0	0	0	0	0	1	0	1
9	Corporate and social responsibility	0	0	0	0	0	0	0	0	1	1
10	Trust building	0	0	0	0	0	0	0	0	0	1

Table 7.4 Final reachability matrix

S. No.	Human dimensions	1	2	3	4	5	6	7	8	9	10	Driver power	Rank
1	Management commitment	1	1	1	1	1	1	1	1	1	1	10	I
2	Communication	0	1	1	1	1	1	1	1	1	1	9	II
3	Training and education	0	0	1	1	1	1	1	1	1	1	8	III
4	Customer relationship management	0	0	0	1	0	1	0	0	0	1	3	VI
5	Information sharing	0	0	0	0	1	0	1	1	0	1	4	V
6	Leadership	0	0	0	0	0	1	1	1	1	1	5	IV
7	Organization culture	0	0	0	0	0	0	1	1	0	1	3	VI
8	Collective learning and chain harmonization	0	0	0	0	0	0	0	1	0	1	2	VII
9	Corporate and social responsibility	0	0	0	0	0	0	0	0	1	1	2	VII
10	Trust building	0	0	0	0	0	0	0	0	0	1	1	VIII
	Dependency	1	2	3	4	4	5	6	7	5	10	47	
	Rank	VIII	VII	VI	V	V	IV	III	II	IV	I		

shows the entries in the matrix with the driver and dependence scores of various dimensions. The last column presents the values of the driving power and last row presents the values of the dependence power of variables.

7.4.4 Level partitions

After the creation of the final reachability matrix, further processing is done to make the hierarchical model based on association among the criteria. To do this, the sets of reachability and antecedent for each dimension are obtained for the identified dimensions and associated levels. The dimensions for which both sets of reachability and sets of intersection are the same are positioned at the top in the ISM hierarchy. The procedure is repeated until all levels of the structural model are determined. In this case, the level identification is completed using six iterations for the nine dimensions. Table 7.5 shows the iterative steps.

Table 7.5 Iterations

Dimensions	Reachability	Antecedent	Intersection	Level
Iteration I				
1	1,2,3,4,5,6,7,8,9,10	1	1	
2	2,3,4,5,6,7,8,9,10	1,2	2	
3	3,4,5,6,7,8,9,10	1,2,3	3	
4	4,6,10	1,2,3,4	4	
5	5,7,8,10	1,2,5	5	
6	6,7,8,9,10	1,2,3,4,6	6	
7	7,8,10	1,2,3,4,5,7,8	7,8	
8	8,10	1,2,3,4,5,6,8	8	
9	9,10	1,2,3,4,5,6,7,9	9	
10	10	1,2,3,4,5,6,7,8,9,10	10	I
Iteration II				
1	1,2,3,4,5,6,7,8,9	1	1	
2	2,3,4,5,6,7,8,9	1,2	2	
3	3,4,5,6,7,8,9	1,2,3	3	
4	4,6	1,2,3,4	4	
5	5,7,8	1,2,5	5	
6	6,7,8,9	1,2,3,4,6	6	
7	7,8	1,2,3,4,5,7,8	7,8	
8	8	1,2,3,4,5,6,8	8	II
9	9	1,2,3,4,5,6,7,9	9	II
Iteration III				
1	1,2,3,4,5,6,7	1	1	
2	2,3,4,5,6,7	1,2	2	
3	3,4,5,6,7	1,2,3	3	
4	4,6	1,2,3,4	4	
5	5,7	1,2,5	5	
6	6,7	1,2,3,4,6	6	
7	7	1,2,3,4,5,7	7	III

(Continued)

Table 7.5 (Continued) Iterations

Dimensions	Reachability	Antecedent	Intersection	Level
Iteration IV				
1	1,2,3,4,5,6	1	1	
2	2,3,4,5,6	1,2	2	
3	3,4,5,6	1,2,3	3	
4	4,6	1,2,3,4	4	
5	5	1,2,5	5	IV
6	6	1,2,3,4,6	6	IV
Iteration V				
1	1,2,3,4	1	1	
2	2,3,4	1,2	2	
3	3,4	1,2,3	3	
4	4	1,2,3,4	4	V
Iteration VI				
1	1,2,3,4	1	1	
2	2,3,4	1,2	2	
3	3, 4	1,2,3	3	VI
Iteration VII				
1	1,2,3,4	1	1	
2	2,3,4	1,2	2	VII
Iteration VIII				
1	1,2,3,4	1	1	VIII

7.4.5 ISM model development

The conical matrix is formulated by joining together risk dimensions which are at a similar level in the corresponding rows and columns of the final reachability matrix (FRM). Thereafter, the number of ones are aggregated in the rows to determine the power of driving and in the same manner the total number of ones are aggregated in the columns up to obtain the power of dependency. Based upon the sums, ranks of dimensions are obtained.

The ISM model as developed from the digraph representing the hierarchy of dimensions is presented in Figure 7.3. Dimension at Level I is positioned at the topmost position in the model followed by dimensions on the second position at Level II, dimensions on the third position at Level III and similarly the variables are placed up to the last level one by one. In the study, two dimensions are placed at Level II and Level III and one each from Level IV to Level VIII. The model shows the dimensions and their positions at various levels and thus depicts the inter-association between the dimensions. From the level partition results, the "process improvement"

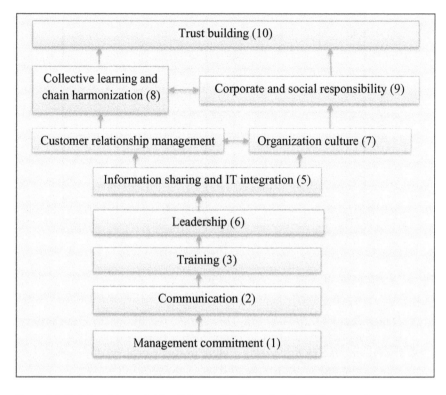

Figure 7.3 Relationship model for SC coordination using the ISM approach.

dimension is positioned at Level I followed by employee participation and corporate social responsibility dimensions at Level II. In a similar manner, the human dimensions are positioned at the determined levels.

7.5 MODEL ANALYSIS AND VALIDATION

To assess the power of dimensions, viz. driver and dependent power and to further classify them under different clusters, namely, autonomous, linkage, dependent and independent, MICMAC analysis as discussed in Section 2.1 is used. Figure 7.4 presents the results of MICMAC analysis which is used to validate the ISM model for vendor selection dimensions.

The four clusters with definitions are discussed as under:

1. *Autonomous dimensions*: Those dimensions possessing the weakest powers of driving and dependence fall in the group of autonomous dimensions.
2. *Linkage dimensions*: The dimensions having the strongest powers of driving and dependence falls in the group of linkage dimensions.

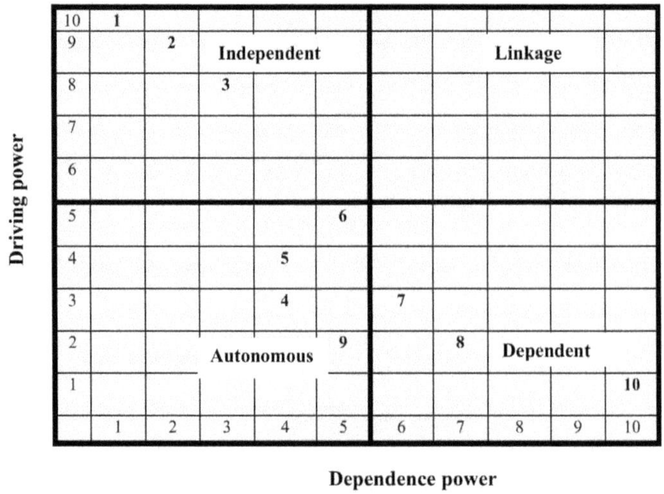

Figure 7.4 MICMAC showing power of dimensions.

3. *Dependent risk dimensions*: This category contains criteria that possess weak driving and strong dependence powers.
4. *Independent risk dimensions*: Here in this category, dimensions possess strong power of driving and weak power of dependence.

The four quadrants in Figure 7.4 present these four groups of dimensions and are briefly discussed as under.

Quadrant 1: In the study, the dimensions, viz. information sharing, customer relationship management, leadership style and corporate and social responsibility are placed under the autonomous category.
Quadrant 2: In the study, three dimensions, viz. organization culture, collective learning and chain harmonization and trust building belong to this cluster.
Quadrant 3: No dimension belongs to this category.
Quadrant 4: In the study, three dimensions, viz. management commitment, communication and co-operation, and training belong to this quadrant.

7.6 CONCLUSION

Supply chain members do not work in isolation as they are dependent on each other to ensure the flow of goods and information. Global offshore outsourcing and advancements in application of informational and communication technology in supply chains have resulted in dependency on each other to increase. Focus by managers on SC human dimensions help them for supply chain coordination in an effective manner among supply chain entities. The dimensions help to build a strong relationship with suppliers. A successful supply chain should be well coordinated which not only enhances supply chain management (SCM) performance by taking into account various administrative and technical dimensions but also human and behavioural dimensions too.

SC human dimensions are critical to improve the SC performance and growth of firms to compete globally. The chapter identifies and develops a hierarchal model of these dimensions that would assist the managers in effective SC coordination. In addition, to analyze the interaction among various SC dimensions, a hierarchal model has been built. The major findings of the developed ISM model are as follows:

- The MICMAC analysis shows that customer relationship management, information sharing, leadership and corporate and social responsibility are autonomous dimensions for improving the SCM performance. Autonomous dimensions having weak driving and weak dependence power are relatively not connected with the model. These dimensions have very less influencing power on the other dimensions of the system.
- The dimensions trust building, collective learning and process improvement are weak drivers but they possess dependence power strong on other dimensions. They are placed at the topmost position in the ISM model.
- No dimension is under linkage dimension in the study.
- The MICMAC diagram shows that dimensions such as management commitment, communication and cooperation and training and education possess strong driving power and weak dependence power and thus they are placed at the lowermost position of the model.

FURTHER READING

Feizabadi, J., Alibakhshi, S. (2022). Synergistic effect of cooperation and coordination to enhance the firm's supply chain adaptability and performance. *Benchmarking: An International Journal*, 29(1), 136–171. https://doi.org/10.1108/BIJ-11-2020-0589.

Kumar, A., Mangla, S.K., Luthra, S., Ishizaka, A. (2019). Evaluating the human resource related soft dimensions in green supply chain management implementation. *Production Planning & Control*, 1–17. https://doi.org/10.1080/09537287.2018.1555342.

Lin, L., Li, T. (2010). An integrated framework for supply chain performance measurement using six-sigma metrics. *Software Quality Journal*, 18(3), 387–406.

Mahdiraji, H.A., Kamardi, A.A., Beheshti, M. (2022). Analysing supply chain coordination mechanisms dealing with repurposing challenges during Covid-19 pandemic in an emerging economy: a multi-layer decision making approach. *Operations Management Research*, 15, 1341–1360.

Tsanos, S., Zografos, K., Harrison, A. (2014). Developing a conceptual model for examining the supply chain relationships between behavioural antecedents of collaboration, integration and performance. *The International Journal of Logistics Management*, 25(3), 418–462. https://doi.org/10.1108/IJLM-02-2012-0005.

Chapter 8

Modeling cybersecurity risk dimensions in digital supply chains operating in an offshore environment

8.1 INTRODUCTION

Leimeister (2010) defines IT outsourcing as handing over some or all information system functions of an organization to one or more third-party vendors. These functions include IT assets, activities, people, processes or services on a mutually agreed contract for a particular period of time. The outsourcing of information technology services is now a well-established practice and is used by companies across the globe. Though IT outsourcing makes enterprises more agile and cost-effective, it is evident that IT outsourcing involves cybersecurity risks. According to a World Bank report, 40% of business firms consider security as a primary obstacle in adopting cloud services because of issues related to data privacy, compliance and access controls, lack of trust and transparency and shared responsibility. These concerns on cybersecurity are more for smaller business firms as they outsource more of their IT needs. According to the Global report by Ponemon Institute (2021), the cybersecurity incidents caused by insiders have increased by 47% since 2018. For instance, a former engineer of amazon web services was found guilty of stealing the personal information of 100 million customers linked to Capital One in 2019. Other examples of IT outsourcing risk are Salesforce's multi-hour cloud meltdown because of error in database which grants the users right to use (May 2019); in June 2019, Google's cloud outage brought down services of YouTube, Gmail and Snapchat in various regions of the United States. Seh et al. (2020) studied data hacking issues related to the healthcare industry and stated that such breaches are a concern for clients, stakeholders, organizations and businesses. The work stated that incidents related to hacking are the dominant forms of risks which are behind the healthcare industry data breaches, followed by prohibited internal releases. The growing capability of cloud-based IT systems have advanced the threats related to cybersecurity, thus making cybersecurity risk concerns dominant in all forms of IT outsourcing business. While outsourcing their IT needs, firms unequivocally presume that IT outsourcing service providers abide by their concerns for risks associated with cybersecurity.

DOI: 10.1201/9781032707884-8

The risk survey conducted globally by Gartner, AXA, Society of actuaries, Deloitte firms in 2018 revealed that cybersecurity and data outages appeared as one of the top risks which modern enterprises face today. According to Xue et al. (2013), Urciuoli (2015), and Urciuolia and Hintsa (2016) the studies in literature does not address the consequences of cyberthreats satisfactorily in the context of supply chain at different levels.

> Information security poses a huge challenge in business as much of our information services are outsourced. However, some important challenges also need to be considered, such as lack of data, insiders, IT vulnerabilities, regulatory frameworks, criminal behaviour, etc. Hence, recommendations are made for managers to improve their understanding of supply chain security.
>
> *(Urciuolia and Hintsa 2016)*

The chapter identifies key cyber risk dimensions and model the association among these dimensions using an interpretive structural modeling (ISM) approach. A theoretical framework consisting of two phases has been used to model the association among the dimensions and analyse these dimensions with respect to their driving and dependence power.

8.2 CYBERSECURITY RISK DIMENSIONS

The main dimensions considered in the study are based on studies undertaken by Bomhard and Daum (2021), Pandey et al. (2020), Ghadge et al. (2019), Urciuoli (2015), and Peck (2006). These dimensions are related to cross-country risk, within-country risk, physical threats, software vulnerabilities, cyber assaults or attacks, insider threats because of humans and contextual risk factors. The various dimensions along with the sub-dimensions are presented in Table 8.1.

8.3 CONCEPTUAL FRAMEWORK FOR CYBERSECURITY RISK DIMENSIONS

The present section presents details of the conceptual framework and methodology used. Figure 8.1 presents the framework. The proposed framework consists of two phases.

- In phase I, various dimensions associated with cybersecurity risk are identified through literature studies and conversation with domain experts. Though IT outsourcing makes enterprises more agile and cost-effective, it is evident that IT outsourcing involves cybersecurity risks. These dimensions are related to cross-country risk,

Table 8.1 Cybersecurity risk dimensions

Dimensions	Sub-dimensions	Description
Physical threats	ICT devices Control mechanisms	These types of risks affect the supply chain functions and security of physical infrastructure components. The physical infrastructure includes switches, servers, routers, firewalls and other ICT devices. For instance, when a natural disaster occurs, the information flow among the entities in the supply chain network is disrupted.
Software vulnerabilities	Outdated subsystems	Executing different tasks entities in the supply chain depends on different software systems. These vulnerabilities are embedded into the software during the design or implementation phase. The vulnerabilities are discovered or updated on a continual basis before making public for intended users. On the other hand, some of the systems are neither updated nor fixed at all. In both cases, they are extremely vulnerable.
Cyber assaults or attacks	Direct attacks Indirect attacks	The risk due to cyberattacks is direct and indirect. Virus and hacking attacks come under direct attacks. They impact operations which consist of risk associated with industry spying or infringement of IPR. Under the indirect type of attacks, the hackers lay out a "bait" to breach the security of systems.
Insider threats because of humans	Deliberate Premeditated	Insider threats are caused by workers in which they use their authorized access to harm the organization. For example, insiders could provide substantial support to external hackers by revealing weak points, or disclosing authenticated passwords. Whether the cyber breach by an employee is deliberate or accidental or negligent, it is termed an insider threat. Thus, the human factor can pose the biggest threat to a firm's cybersecurity.
Within-country risk	Government regulations Legal standards	Within-country risk sources involve regulatory rights or government policies on digital connectivity, digital e-commerce and digital intellectual property. It also entails riders against foreign companies towards digital connectivity. Legal standards that prevent internet frauds fall under within-country risk category.
Cross-country risk	Geopolitics Bilateral relations	Cross-country or geopolitics risk exacerbates when bilateral relations between home and host country worsens. Cross-country or geopolitics risk exacerbates when bilateral relations between home and host country worsens because of divergent ICT standards between economies (United States vs China norms) related to digitization systems. The cross-country risks pose implications for national security. Moreover, the digital supply chains make firms more prone to cyber risks.
Contextual risks	Procurement Staffing Funding Training IT infrastructure	The contextual risks are associated with supporting processes such as procurement, staffing, funding, training and development. The digital infrastructure (automated processes, strong AI-supported algorithms and cloud computing services).

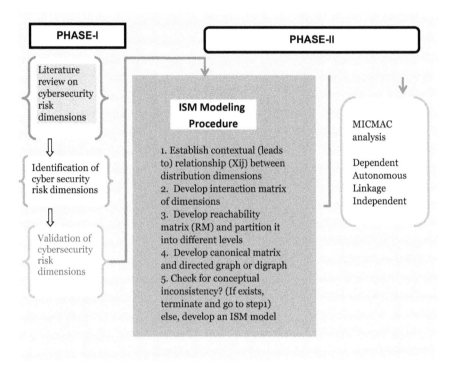

Figure 8.1 Theoretical outline for cybersecurity risk dimensions.

within-country risk, physical threats, software vulnerabilities, cyber assaults or attacks, insider threats because of humans and contextual risk factors as shown in Table 8.1.

- In phase II, ISM approach is used to model the complex relationship among cybersecurity dimensions. The various procedural steps in the ISM approach are: (i) establishment of contextual (leads to) relationship (Xij) between distribution dimensions, (ii) development of interaction matrix of dimensions, (iii) development of the reachability matrix (RM) and partitioning it into different levels, (iv) development of canonical matrix and directed graph or digraph, and (v) checking for conceptual inconsistency? (if exists, terminate and go to step 1), else develop an ISM model.

- To study the dynamics of vendor selection criteria a MICMAC diagram is constructed to examine the driver and dependence power of dimensions. This is accomplished by portraying these dimensions under four groups, i.e., dependent, independent, linkage and autonomous.

8.4 HIERARCHAL STRUCTURAL MODEL FOR CYBERSECURITY RISK DIMENSIONS

8.4.1 Structural self-interaction (SSI) matrix

The relative association among the dimensions is represented as "lead to", which states that how one dimension leads to another dimension. Consultations with a total of six experts, out of which four were from the companies and two were from academic units were made to formulate the contextual relation among the upstream dimensions. On the basis of responses, the SSI matrix is made as presented in Table 8.2. The matrix was formed using V, A, X, and O symbols.

8.4.2 Reachability matrix

The SSI matrix is converted into a (0, 1) matrix, known as the initial reachability matrix (IRM), as shown in Table 8.3. The symbols representing the relationship among the dimensions in Table 8.2 are substituted by 0 and 1 by following the rules:

  i. If (i, j) in the SSI matrix is shown by V, then (i, j) in the RM will be 1 and (j, i) is 0.
  ii. If (i, j) in the SSI matrix is shown by A, then (i, j) in the RM will be 0 and (j, i) is 1.
  iii. If (i, j) in the SSI matrix is shown by X, then both (i, j) and (j, i) in the RM will be 1.
  iv. If (i, j) in the SSI matrix is shown by O, then both (i, j) and (j, i) in the RM is 0.

8.4.3 Development of the final reachability matrix

The transitive relation between the dimensions is eliminated, and RM is obtained in the final from the initial one. Table 8.4 shows the entries in

Table 8.2 Self-interaction matrix for dimensions

S. No.	Dimensions	2	3	4	5	6	7
1	Within-country risk	A	O	O	A	V	O
2	Software vulnerabilities		O	A	A	O	V
3	Contextual risk			A	O	V	V
4	Insider threats because of humans				O	O	V
5	Physical threats					O	V
6	Cross-country risk						A
7	Cyber assaults or attacks						

Table 8.3 Reachability matrix

Dimensions	1	2	3	4	5	6	7	Driving power
1 Within-country risk	1	0	0	0	0	1	0	2
2 Software vulnerabilities	1	1	0	0	0	0	1	3
3 Contextual risk	0	0	1	0	0	1	1	3
4 Insider threats because of humans	0	1	1	1	0	0	1	4
5 Physical threats	1	1	0	0	1	0	1	4
6 Cross-country risk	0	0	0	0	0	1	0	1
7 Cyber assaults or attacks	0	0	0	0	0	1	1	2
Dependence power	3	3	2	1	1	4	5	

Table 8.4 Reachability matrix with transitive links

Dimensions	1	2	3	4	5	6	7	Driving power
1 Within-country risk	1	0	0	0	0	1	0	2
2 Software vulnerabilities	1	1	0	0	0	1*	1	4
3 Contextual risk	0	0	1	0	0	1	1	3
4 Insider threats because of humans	1*	1	1	1	0	1*	1	6
5 Physical threats	1	1	0	0	1	1*	1	5
6 Cross-country risk	0	0	0	0	0	1	0	1
7 Cyber assaults or attacks	0	0	0	0	0	1	1	2
Dependence power	4	3	2	1	1	7	5	

the matrix with the driver and dependence scores of various dimensions. The last column presents the values of the driving power and last row presents the values of the dependence power of variables. In Table 8.4, to show the transitivity, 1* is used for the dimensions which follow it. The final reachability matrix so obtained after a transitivity check is shown in Table 8.4.

8.4.4 Level partition

After creation of the final reachability matrix, further processing is done to make the hierarchical model based on association among the criteria. To do this, the sets of reachability and antecedent for each dimension are obtained for the identified dimensions and associated levels. The dimensions for which both sets of reachability and sets of intersection are the same are positioned at the top in the ISM hierarchy. The procedure is repeated until all levels of the structural model are determined. In this case, the level identification is completed using four iterations as shown in Table 8.5 for the seven dimensions.

Table 8.5 Iterations

Dimension	Reachability	Antecedent	Intersection	Level
Iteration I				
1	1,6	1,2,3,4,5	1	
2	1,2,6,7	2,4,5	2	
3	3,6,7	3,4	3	
4	1,2,3,4,6,7	4	4	
5	1,2,5,6,7	5	5	
6	6	1,2,3,4,5,6,7	6	I
7	6,7	2,3,4,5,7	6	
Iteration-II				
1	1	1,2,3,4,5	1	II
2	1,2,7	2,4,5	2	
3	3,7	3,4	3	
4	1,2,3,4,7	4	4	
5	1,2,5,7	5	5	
7	7	2,3,4,5,7	6	II
Iteration III				
2	2	2,4,5	2	III
3	3	3,4	3	III
4	2,3,4	4	4	
5	2,5	5	5	
Iteration IV				
4	4	4	4	IV
5	5	5	5	IV

8.4.5 Formulation of the conical matrix

The conical matrix is formulated by joining together risk dimensions which are at a similar level in the corresponding rows and columns of the FRM. Table 8.6 shows the elements in the conical matrix. Thereafter, the total number of ones are aggregated in the rows to determine the power of driving and in the same manner the total number of ones are aggregated in the columns up to obtain the power of dependency. Based upon the sums, ranks of dimensions are obtained.

8.4.6 ISM model development

The ISM model as developed from the digraph representing the hierarchy of dimensions is presented in Figure 8.2. Dimension cross-country risk at Level I is placed at the topmost position in the model followed by dimensions within

Table 8.6 Conical matrix

Dimensions	6	I	7	2	3	4	5	Driving power	Level
6	I	0	0	0	0	0	0	I	I
I	I	I	0	0	0	0	0	2	II
7	I	0	I	0	0	0	0	2	II
2	I*	I	I	I	0	0	0	4	III
3	I	0	I	0	I	0	0	3	III
4	I*	I*	I	I	I	I	0	6	IV
5	I*	I	I	I	0	0	I	5	IV
Dependence power	7	4	5	3	2	I	I		

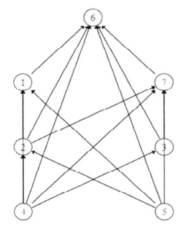

1. Within-country risk
2. Software vulnerabilities
3. Contextual risk
4. Insider threats because of human
5. Physical threats
6. Cross-country risk
7. Cyber Assaults or Attacks

Figure 8.2 Digraph of dimensions.

country and cyberattacks (both direct and indirect attacks) on second position at Level II, dimensions, viz. system breakdown and contextual risk on third position at Level III and finally dimensions insider threats and physical threats which disrupt the supply chain's functions are placed at the bottom level. These dimensions which occur at the lowermost level in the ISM model drive all other dimensions placed in the model. Figure 8.3 presents the ISM model.

8.5 MODEL ANALYSIS AND VALIDATION

To assess the power of dimensions, viz. driver and dependent power and to further classify them under different clusters, namely, autonomous, linkage, dependent and independent, MICMAC analysis as discussed in Section 2.1

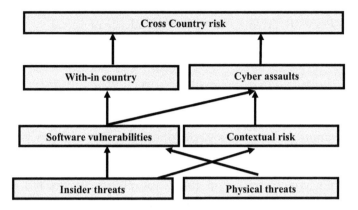

Figure 8.3 ISM model.

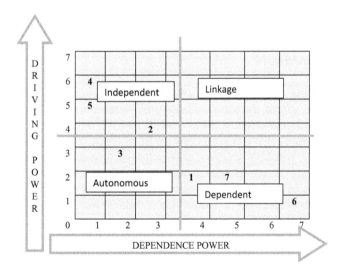

Figure 8.4 MICMAC diagram of dimensions.

is used. Figure 8.4 presents the results of MICMAC analysis which is used to validate the ISM model for vendor selection dimensions.

The four clusters with definitions are discussed as under:

1. *Autonomous dimensions*: Those dimensions possessing the weakest powers of driving and dependence fall in the group of autonomous dimensions.
2. *Linkage dimensions*: The dimensions having the strongest powers of driving and dependence falls in the group of linkage dimensions.

3. *Dependent risk dimensions*: This category contains criteria which possess weak driving and strong dependence powers.
4. *Independent risk dimensions*: Here in this category, dimensions possess strong power of driving and weak power of dependence.

The four quadrants in Figure 8.4 present these four groups of dimensions and are briefly discussed as under.

- *Quadrant 1*: In the study, the dimension contextual risk is placed under the autonomous category.
- *Quadrant 2*: In the study, three dimensions, viz. cross-country, with-in country risk and cyber assaults or attacks fall into this quadrant.
- *Quadrant 3*: In the study, under this quadrant no dimension is found.
- *Quadrant 4*: In the study, dimensions software vulnerabilities, insider threats because of human and physical threats are placed in this quadrant.

8.6 CONCLUSION

The outsourcing of IT services is now a well-established practice and is used by companies across the globe. Though IT outsourcing makes enterprises more agile and cost-effective, it is evident that IT outsourcing contributes towards cybersecurity risks. According to a World Bank report, 40% of business firms consider security as a primary obstacle in adopting cloud services because of issues related to data privacy, compliance and access controls, lack of trust and transparency and shared responsibility. These concerns on cybersecurity are more for smaller business firms as they outsource more of their IT needs.

This chapter describes various dimensions related to cybersecurity breaches in digital supply chains because of IT offshoring. The main dimensions considered in the study are based on studies undertaken by Peck (2006), Urciuoli and Hintsa (2016) and Ghadge et al. (2019). These dimensions are related to physical threats, software vulnerabilities, cyber assaults or attacks, insider threats because of humans, cross-country risk, within-country risk and contextual risk. The main question answered in this chapter is: *What is the association among these dimensions which affect IT offshoring at the level of supply chains?* To answer this question, a theoretical framework has been developed using the ISM approach which helps to analyse the structural association among the dimensions affecting IT offshoring. In the ISM model, the dimension cross-country risk is positioned at the top level followed by dimensions within country and cyberattacks (both direct and indirect attacks) on second position at Level II. The dimensions, viz. system breakdown and contextual risk are on the third position at Level III and finally dimensions insider threats because of human and physical threats are placed at the bottom level. Results of MICMAC analysis depict the dynamics of these dimensions with respect to their powers of driving and dependence.

FURTHER READING

Cebula, J.J., Young, L.R. (2010). A taxonomy of operational cyber security risks. Technical Note CMU/SEI-2010-TN-028, Software Engineering Institute, Carnegie Mellon University..

Cha, S.Y. (2022). The Art of Cyber Security in the Age of the Digital Supply Chain: Detecting and Defending Against Vulnerabilities in Your Supply Chain. In: MacCarthy, B.L., Ivanov, D. (eds) *The Digital Supply Chain* (pp. 215–233). Elsevier, ISBN 9780323916141.

Ghadge, A., Weiß, M., Caldwell, N.D., Wilding, R. (2019). Managing cyber risk in supply chains: a review and research agenda. *Supply Chain Management: An International Journal*, 25(2), 223–240. https://doi.org/10.1108/scm-10-2018-0357.

Hodosi, G., Rusu, L. (2019). *IT Outsourcing: Definition, Importance, Trends and Research* (pp. 1–9). SpringerBriefs in Information Systems, Springer, Cham. https://doi.org/10.1007/978-3-030-05925-5_1.

IBM Security. (2021). Cost of insider threats global report. Technical report, Ponemon Institute.

ICAEW Insights. (2022). How to manage the cyber security risks lurking within supply chains Available at: https://www.icaew.com/insights/view-points-on-the-news/2022/oct-2022 [Accessed on 27 June 2023].

Jüttner, U., Peck, H., Christopher, M. (2003). Supply chain risk management: outlining an agenda for future research. *International Journal of Logistics Research and Applications*, 6(4), 197–210.

Pandey, S., Singh, R.K., Gunasekaran, A., Kaushik, A. (2020). Cyber security risks in globalized supply chains: conceptual framework. *Journal of Global Operations and Strategic Sourcing*, 13(1), 103–128. https://doi.org/10.1108/JGOSS-05-2019-0042.

Peck, H. (2006). Reconciling supply chain vulnerability, risk and supply chain management. *International Journal of Logistics Research and Applications*, 9(2), 127–142.

Sindhuja, P. (2014). Impact of information security initiatives on supply chain performance an empirical investigation. *Information Management and Computer Security*, 22(5), 450–473.

Urciuoli, L. (2015) Cyber-resilience: a strategic approach for supply chain management. *Technology Innovation Management Review*, 5(4), 13–18.

Urciuoli, L., Hintsa, J. (2016). Adapting supply chain management strategies to security -an analysis of existing gaps and recommendations for improvement. *International Journal of Logistics Research and Applications*, 20(3), 276–295.

Xue, L., Zhang, C., Ling, H., Zhao, X. (2013). Risk mitigation in supply chain digitization: system modularity and information technology governance. *Journal of Management Information Systems*, 30(1), 325–325.

Chapter 9

Risk dimensions disrupting the automotive supply chain

Covid-19 pandemic – case study

9.1 INTRODUCTION

Over the years, the supply chain in the automotive sector has undergone major transformations with development into a complex multi-national distribution network. Both manufacturing and assembly functions mostly rely upon raw materials and components and that are sourced from all corners of the world, which results in reliance on third-party vendors. In this delicate supply chain network, a single disturbance can halt or jeopardize the entire supply chain process.

The research problems related to supply chain risk in Covid-19 includes critical scrutiny of risks which includes internal risks (arising within) or external risks arising outside the organization. In addition, it is pertinent to outline the nature of these risks based on types, viz. physical, financial, information and operational risks. Based upon the recent literature studies, the chapter recognizes 24 probable failure modes affecting the supply chains of automobile companies. The identified failure modes are related to suppliers, logistics and material handling, warehouse and distribution centre, manufacturer, customer covering strategy, operations, and product-related issues in the supply chain. For the failure modes, the score of probability of occurrence (O) and severity (S) detection on a 1–10 scale is determined to calculate the risk output, i.e., the RPN. For example, in the failure mode and effects analysis (FMEA) analysis under Cluster I, i.e., supplier issues, five potential failure modes, viz. quality problems, end customer, raw material shortages, economic issues and regulatory issues are identified. Of these five, the failure mode supplier quality has high weighted risk priority number, 5.75, followed by economic position weighted risk priority number, 4.72, respectively. Similarly, the FMEA analysis for other dimensions, viz. logistics and material handling, manufacturer and customer is carried out.

Of the 24 failures modes, 12 are finally chosen and structural association among them is modelled using the interpretive structural modeling (ISM) approach. The model portrays the hierarchy of the risk dimensions. The driver-dependence power is examined using the Matrice d'Impacts Croisés Multiplication Applied to a Classification (MICMAC) diagram.

　　　　　　　　　　　　DOI: 10.1201/9781032707884-9

The MICMAC diagram classifies these risks into four distinct clusters, viz. autonomous, dependent, independent and linkage. All the risk variables are placed on the basis of their powers of driving and dependency into four quadrants.

9.2 RISKS DIMENSIONS THAT DISRUPT THE SUPPLY CHAIN OF AUTOMOBILE COMPANIES

The risk dimensions that disrupt the automotive supply chain are summarized through the critical scrutiny of literature studies. The 12 dimensions are equipment failures, financial risk, poor planning and scheduling, labour relations, quality, inventory risk, technology, environmental and regulatory risk, outsourcing, inaccurate forecast and logistics risk and cybersecurity (years 2013–2023) and discussion with experts. Table 9.1 presents the definitions and cited literature.

9.3 CONCEPTUAL FRAMEWORK FOR RISK MITIGATION IN THE AUTOMOTIVE SUPPLY CHAIN

The two-stage framework has been developed to examine and model the various risk sources related with the supply chain of automotive companies presented in Figure 9.1 Based on strategy, operation, and product-related issues in the supply chain, the failure modes are broadly grouped into five categories, viz.:

   i. Vendors
   ii. Distribution logistics and material handling
   iii. Warehouse and distribution centre
   iv. Manufacturer, and
   v. Customer

The influence of various failure modes on automotive supply chain is ascertained and presented in Table 9.3. Subsequently, on the basis of the score of probability of occurrence (O) and severity (S) detection on a 1–10 scale, the output, i.e., the RPN is determined for all the failure modes under each of the clusters. Finally, in stage II of the framework, 12 failure modes with high RPN number are selected and the association between the risk dimensions is analysed using the ISM approach. Also, the MICMAC diagram showing the driver-dependence relationship among risk dimensions is constructed and dynamics of these dimensions is studied with respect to the automotive supply chain of the tractor company. Both the stages are shown in Figure 9.1 and are discussed in Sections 9.3.1 and 9.3.2, respectively.

Table 9.1 Risk dimensions that disrupt the automotive supply chain during the Covid-19 pandemic

S. No.	Risk dimensions	Literature studies	Description
1	Equipment failures	Dehdar et al. (2018), Gurtu and Johny (2021), Sharma (2022), Rosangela Maria Vanalle et al. (2020), Duhamel et al. (2016), Wang et al. (2017)	Equipment or machine failure leads to delay in production which ultimately results in loss of orders from the customer.
2	Financial risk	Errico et al. (2022), Sharma (2022), Dhone and Kamble (2015), Gupta and Singh (2015)	Financial risk arises because of a variety of reasons, including fluctuation in market and unexpected costs.
3	Poor planning and scheduling	Surange and Bokad (2022), Dehdar et al. (2018), Rosangela Maria Vanalle et al. (2020), Duhamel et al. (2016), Wang et al. (2017), Gupta and Singh (2015)	Lack of project management results in manufacturing delays, increase in production costs and wastage of man, machine, material, etc.
4	Labour relations	Gurtu and Johny (2021), Dehdar et al. (2018), Duhamel et al. (2016), Wang et al. (2017); Salehi Heidari et al. (2018), Dhone and Kamble (2015)	Owing to high work load and fatigue, labour problems may occur which may result in non-fulfilment of orders timely and disrupts the supply chain.
5	Quality	Dehdar et al. (2018), Kumar Sharma and Bhat (2014), Rosangela Maria Vanalle et al. (2020), Dhone and Kamble (2015), Gupta and Singh (2015)	Risk related to quality arises because of improper manufacturing, supplier failure, changes in product design and issues with technical reliability.
6	Inventory risk	Gurtu and Johny (2021), Duhamel et al. (2016), Wang et al. (2017), Salehi Heidari et al. (2018), Sharma (2022)	Inventory risks are related to scarcity of raw material or procurement issues and uncertainty in demand which affects the production schedules.
7	Technology	Surange and Bokade (2022), Gurtu and Johny (2021), Sharma (2022); Salehi Heidari et al. (2018), Kumar Sharma and Bhat (2014), Sawik (2013)	Technology risk arises with inadequate use of latest technology and less investment in new infrastructure and machinery.

(*Continued*)

Table 9.1 (Continued) Risk dimensions that disrupt the automotive supply chain during the Covid-19 pandemic

S. No.	Risk dimensions	Literature studies	Description
8	Environmental and regulatory risk	Gurtu and Johny (2021), Sharma (2022), Rosangela Maria Vanalle et al. (2020), Salehi Heidari et al. (2018), Dhone and Kamble (2015), Gupta and Singh (2015)	Environmental and regulatory risk risks are present throughout the lifecycle of a product. Not only limited to greenhouse gas emissions but also air quality, land use, toxic wastes, deforestation and energy use are all important criteria.
9	Offshore outsourcing	Surange and Bokade (2022), Rosangela Maria Vanalle et al. (2020), Dhone and Kamble (2015), Gupta and Singh (2015), Kumar Sharma and Bhat (2014)	When parties deviate from the contract or agreement, outsourcing risk arises in which supplier fails to meet the timeline with respect to shipment of orders.
10	Inaccurate forecast	Katsaliaki et al. (2022), Gurtu and Johny (2021), Dehdar et al. (2018), Rosangela Maria Vanalle et al. (2020), Duhamel et al. (2016), Wang et al. (2017), Sawik (2013)	In appropriate forecasting technique, and use of improper data may result in inaccurate forecasts. Most common forecast errors are bias and mean absolute deviation.
11	Distribution logistics	Surange and Bokade (2022), Rosangela Maria Vanalle et al. (2020), Duhamel et al. (2016), Wang et al. (2017), Dhone and Kamble (2015), Gupta and Singh (2015)	Logistics risk occurs because of delayed deliveries, more transportation cost and more response time. It includes both in-bound and outbound logistic operations.
12	Cybersecurity	Sang Yoon Cha (2022), Pandey et al. (2020), Urciuoli and Hintsa (2016), Xue et al. (2013), Jüttner et al. (2003)	In a digitally connected world, data security in the supply chain has become a major challenge in the adoption of technology among the business houses because of the threats related to data breaches, cyberattacks and malware.

STAGE-I

FAILURE MODE AND EFFECTS ANALYSIS

1. Preparation of information related to subject

2. Determine potential failure modes

3. Check the potential effects of failure

4. Find severity ranking

5. Determine the potential causes of failure

6. Determine probability of occurrence

7. List detection/current control

8. Determine detection score

9. Determine weight & calculate risk priority number

10. Suggest corrective and remedial actions

11. Implement Modifications

12. Prepare FMEA Report

STAGE-II

Prepare a list of potential risk sources in automotive supply chain network

1. Establish contextual (leads to) relationship (Xij) between distribution dimensions

2. Develop interaction matrix of dimensions

3. Develop reachability matrix (RM) and partition it into different levels

4. Develop canonical matrix and directed graph or digraph

5. Check for conceptual inconsistency? (If exists, terminate and go to step1) else, develop an ISM model

6. Represents association in to model for risks affecting supply chain network

Figure 9.1 Two-stage conceptual framework for automotive supply chain risk.

9.3.1 Overview of the FMEA analysis

The FMEA as a tool was first used by the NASA in the year 1963 as a formal design methodology to meet their dependability requirements. Since 1963, the tool has been widely used for accomplishing safety and reliability goals of products and processes in military, nuclear and automotive industries. The FMEA is a systematic procedure which helps to analyse the system on the basis of possible failure modes, their causes and effects. The FMEA is performed in initial stages of the product development cycle in order to mitigate the potential failure effects in later stages. As shown in Figure 9.1, the FMEA is conducted by following the steps as listed below:

1. Identification of the system/subsystem to be examined.
2. Constructing the block diagram to identify relations among the components of the system.
3. Determining the failure modes, their causes and their effects on the sub-systems.
4. Examining the failure modes for their severity and finding the severity score (S).
5. Identification of failure detection methods and finding the detection score (D).
6. Estimating the occurrence probability (O) and score of failure modes.
7. Calculation of risk priority number using the product of occurrence, severity and detection.
8. Deciding for correction if required which depends upon the risk priority number.
9. Recommending improvements or actions to improve the system performance.
10. Preparation of the FMEA report in tabular format as shown in Table 9.2.

9.3.2 Overview of the ISM approach

Figure 9.1 shows the conceptual framework for automotive supply chain risk. The different steps shown in the figure are discussed as under.

Step 1: Determining the dimensions of interest pertinent to the problem by critical scrutiny of literature or using the group problem-solving technique.
Step 2: Developing a relative association between the recognized dimensions according to which various pairs of dimensions need to be analysed.
Step 3: Development of the SSI (structural self-interaction) matrix to determine pair-wise relations between the dimensions.

Table 9.2 FMEA inputs and output table

Item/ function	Potential failure modes	Potential failure effect(s)	Severity (S)	Potential cause(s) of failure	Occurrence (O)	Detection score (D)	RPN = O×S×D

Step 4: Developing a reachability matrix from the SSI matrix, and checking for transitivity relations. Transitivity refers to the assumption that, if a dimension "*A*" is influencing "*B*", and "*B*" is influencing "*C*", then "*A*" will be necessarily influencing "*C*".

Note: The SSI matrix is changed into a 0, 1 matrix called the binary matrix by changing symbols *V*, *A*, *X* and *O* with 1 and 0 on the basis of rules:

i. If (i, j) value in the SSI matrix is represented by *V*, then (i, j) value in the RM will be 1 and (j, i) value is 0.

ii. If (i, j) value in the SSI matrix is represented by *A*, then (i, j) value in the RM will be 0 and (j, i) value is 1.

iii. If (i, j) value in the SSI matrix is represented by *X*, then both (i, j) and (j, i) values in the RM will be 1.

iv. If (i, j) value in the SSI matrix is represented by *O*, both (i, j) and (j, i) entries in the RM is 0.

Step 5: The reachability matrix is split into different levels.

Step 6: A directed graph called diagraph is drawn based on the relations in the reachability matrix, and links with transitive relations are removed.

Step 7: In this step, the digraph is changed to a hierarchy model ISM by replacement of nodes with dimension statement.

Step 8: In the last step, the ISM model is examined for conceptual inconsistences, if any.

9.4 ILLUSTRATIVE CASE FOR FRAMEWORK IMPLEMENTATION

The Covid-19 pandemic has had an impact on the supply chain of automotive industries. A case of Agricultural Tractor and Farm Company is considered for illustration of the framework. The installed capacity of the company is 12000 tractors in 1 year. The company faced many challenges during the Covid-19 pandemic related to raw material or semi-finished supplies from its suppliers across the globe, particularly to Southeast Asian and African countries. Its production unit is equipped with modern cost of non-conformance (CNC) machines, for machining cylinder heads, blocks, transmission housings, hydraulic and fly wheels housings, axle tubes, cages, timing cases, etc. The company depends upon various vendors for engine components, spares and sub-assemblies.

The network of various entities involved in supply chain network is shown in Figure 9.2. The company procures supplies of raw material and components/parts/their sub-assemblies from suppliers, i.e., *S1, S2, S3, S4...*, logistics is handled by third parties say *D1, D2, D3* and *D4*, respectively.

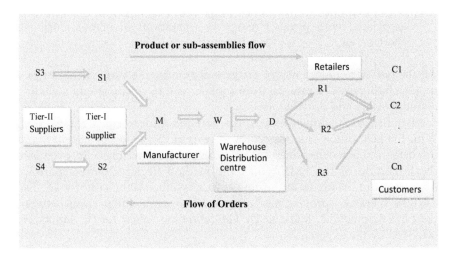

Figure 9.2 Entities in automotive supply chain.

At the plant, manufacturing/assembly is being carried out and the units are delivered to different warehouse locations. Based upon the demand/order size, final shipping is done to distributors and retailers, who further based upon order size deliver the product to the end customer.

The failure mode with respect to various supply chain entities grouped into clusters of suppliers, logistics and material handling, manufacturer and customer based upon operations management, market, business and product issues (Cousins et al. 2004; Singhal et al. 2011; Luthra et al. 2015; Ghadge et al. 2019; Pandey et al. 2020) are defined as under:

- *Operations management issues*: The issues that disrupt the functions or interrupt the supply chain flows related to material and information. For example, equipment failures, financial risk, planning and scheduling, technology, capacity constraints, etc.
- *Market-related matters*: These are random variations of market which change over time. For instance, customer behaviour, customer anticipations and complaints, environmental concerns, quality, price variation, etc.
- *Business issues*: Business issues are strategic issues which undesirably affect the performance of the supply chain. For example, outsourcing decisions, distribution centre and its location, inventory and collaboration, etc., are some of the issues which may affect business or strategic decisions.
- *Product matters*: Such matters are related to the product and product portfolios. Their availability, shortage or back orders, etc. result

in vulnerability of the supply chains. For instance, complexities associated with product design or operations in manufacturing a product, etc.

All the potential failure modes are grouped under the various entities in a supply chain network ranging from supplier end to customer end including manufacturers and distributors.

After listing the failure modes, the potential failure effect is tabulated. With the help of experts, the scores of severity (S), its occurrence (O) and the detection are determined to calculate the RPN number. To overcome the limitation of the FMEA, i.e., the method neglects the comparative importance of the inputs as the importance assumed is the same for all. The analytic hierarchy process method is used to make comparison between the FMEA inputs i.e. S,O and D. The following values for the inputs are provided by the experts:

$$O_{w1} \; Vs \; S_{w2} = 60:40; \quad S_{w2} \; Vs \; D_{w3} = 30:70; \quad \text{and} \quad D_{w3} \; Vs \; O_{w1} = 60:40$$

These values are used to determine weighting factors to conduct the analytical hierarchy process (AHP) procedure. The weight coefficient's $\beta_O = 0.21$, $\beta_S = 0.48$ and $\beta_D = 0.31$ are used in Equation 9.1 to calculate the weighted risk priority number.

$$WRPN = (S \times \beta_S) + (O \times \beta_O) + (D \times \beta_D) \tag{9.1}$$

Table 9.3 presents the details of the FMEA process applied to supply chain risks involved with the automotive sector.

From Table 9.3, it is observed that:

- *Cluster 1*: This cluster presents six failure modes, viz. quality problems, end customer, raw material shortages, and economic regulatory. Of these six the failure modes, vendor quality has high weighted risk priority number 5.75 followed by economic position weighted risk priority number, i.e., 4.72.
- *Cluster 2*: The cluster presents six failure modes, viz. transportation risk, human error, environmental and regulatory risk, labour relations, warehouse/distribution centres and inventory. Of these, transportation risk has high weighted risk priority number, i.e., 7.64 followed by labour relations with weighted risk priority number, i.e., 5.71. As during the Covid-19 pandemic, the problems with transportation modes and labour shortages (both skilled and unskilled manpower) were experienced by the companies which result in these risks with high weighted risk priority number (WRPN) scores.
- *Cluster 3*: This cluster presents two distinct failure modes, viz. inventory inaccuracy/stock control and loss of contract or agreement with

Table 9.3 FMEA table for risk dimensions in automotive SC network

Supply chain entities	Possible modes failure	Possible failure effect	Severity	Possible failure cause	Occurrence	Detection	Detection score	RPN value
Vendor	Quality problems	Manufacturing related quality issues	7	Vendor issues	4	Mechanism for vendor quality control	5	5.75
	End customer	Supply interruption	5	Irregular orders	4	Regular periodic orders, foster coordination among partners	3	4.17
	Raw material shortages	Delays in supply, higher cost	4	Inventory planning issues	4	Forecast and information handling and processing	5	4.31
	Negligent behaviour	Interruptions and resulting delays	6	Careless attitude	3	Not able to fulfil orders	3	4.44
	Economic	Currency fluctuations	4	Instability in demand and prices	3	Changing labour costs and inflationary pressures	7	4.72
	Regulatory	Problems with outsourcing/offshoring	4	Non-compliance with regulatory requirements	3	Control mechanism	3	3.48
Distribution logistics & material handling	Transportation risk	Long delivery cycle and high logistics cost	9	Routing problems and travel modes	4	Vehicle route planning, mode selection	8	7.64
	Human error	Delays, safety issues	6	Mental tension, lack skills, training	3	Training, shift breaks, motivation	3	4.44

(Continued)

Table 9.3 (Continued) FMEA table for risk dimensions in automotive SC network

Supply chain entities	Possible modes failure	Possible failure effect	Severity	Possible failure cause	Occurrence	Detection	Detection score	RPN value
	Environmental and regulatory risk	Accidents, damage of product/vehicle	7	Environment, safety issues	3	Insurance cover, safety, worker training, routine maintenance	4	5.23
	Labour relations	Labour cost, work loss, accidents	8	Employer–employee relations	3	Reward system, ergonomics design work environment	4	5.71
	Warehouse/ distribution centres	Congestion, workforce labour, accidents	4	Ineffective warehouse/ distribution management policy	3	Warehouse/ distribution layout design and stock control policy	4	3.79
	Inventory risk	Inaccurate forecast, uncertain demand/order	4	Inadequate visibility, capacity issues, poor planning	3	CPFR system, barcode MRP and reorder point	7	4.72
Warehouse and distribution centre	Inventory inaccuracy/ stock control	Order fulfilment risk	4	Shortages/ Non-availability	3	Information or order management	6	4.41
	Loss of contract or agreement with 3PL	Supplies held up in transit	7	Opportunistic behaviour	4	Collaboration mechanism	5	5.75
Manufacturer	Inaccurate demand forecast	Inventory issues, uncertain supply and demand	8	Forecasting methods, inappropriate data	4	Application of demand forecast method	6	6.54

(Continued)

Table 9.3 (Continued) FMEA table for risk dimensions in automotive SC network

Supply chain entities	Possible modes failure	Possible failure effect	Severity	Possible failure cause	Occurrence	Detection	Detection score	RPN value
	Capacity limitation	Implementation hurdles, production delay	4	Operator strike, job dissatisfaction	3	Not able to meet orders	4	3.79
	Offshore outsourcing	Vendor delays, high cost and unavailability	9	Improper planning and scheduling, Bullwhip effect, contracts dispute	5	Forecasting, coordination, collaboration, and information processing	7	7.54
	Equipment failure	Production delays, lost orders	7	Maintenance failures, reliability issues, unstable process	4	Shop floor, old technology, preventive maintenance process out of control	5	5.75
	Financial risk	Project delays, orders not fulfilled	8	Reduced motivation, inadequate knowledge, overburden shifts, stress	5	Workplace work environment conditions below standards	7	7.06
	Project management	Manufacturing delays in, high overheads costs, material waste, etc.	8	Flaws in production schedule and production control	3	MPS, work order control proper planning and execution	6	6.33

(Continued)

Table 9.3 (Continued) FMEA table for risk dimensions in automotive SC network

Supply chain entities	Possible modes failure	Possible failure effect	Severity	Possible failure cause	Detection	Occurrence	Detection score	RPN value
	Technology	Quality problems, returns, rejections increased work time, high manufacturing cost	8	Outdated technology used and machinery, equipment obsolete	Investing in new plant and equipment machines	4	6	6.54
Customer	Order management	Backorders or incomplete delayed order	4	Shortages, demand planning inadequate, production issues	Demand forecast, product survey, stock control	3	6	4.41
	Customer complaints	Quality, reliability issues, operational problems	6	Volatile markets, competitors, technology	After sales service, research and development, maintenance policy	4	6	5.58
	Cybersecurity	Cyber assaults, direct and indirect attacks	8	Software vulnerability	Digital transformation digital infrastructure (automated processes, strong AI-supported algorithms)	5	4	6.13

third-party logistics (3PL) service providers. The dimension loss of contract or agreement with 3PL service providers has offshore outsourcing risk weighted risk priority number highest, i.e., 5.75 followed by inventory inaccuracy/stock control weighted risk priority number 4.41.

- *Cluster 4*: This cluster presents seven distinct failure modes, viz. inaccurate forecast, capacity constraints, offshore outsourcing, equipment failure, financial risk, project management and technology risk. The dimension manufacturer has offshore outsourcing risk weighted risk priority number highest, i.e., 7.54 followed by financial risk weighted risk priority number 7.06.
- *Cluster 5*: In this cluster, three failure modes are identified, viz. order fulfilment, customer complaints and cybersecurity. Of these three, the failure mode cybersecurity has a high weighted risk priority number, i.e., 6.13 followed by customer complaints weighted risk priority number, i.e., 5.58. In a digitally connected world, data security has become a major challenge in the adoption of technology among the business houses because of the threats related to data breaches, cyberattacks and malware. To counter these threats, the demand for cybersecurity solutions has increased. According to fortune business insights, the global cybersecurity market is expected to reach $376.32 billion by the year 2029.

Table 9.4 Interaction matrix for dimensions

Dimensions	Dimensions										
	12	11	10	9	8	7	6	5	4	3	2
Offshore out sourcing risk	O	X	A	O	O	A	A	O	V	A	O
Quality	O	A	A	A	A	A	O	O	O	A	
Inaccurate demand forecast	O	A	O	V	O	O	O	O	V		
Inventory risk	O	O	O	A	A	O	O	O			
Financial risk	O	V	O	V	O	O	O				
Transportation risk	X	V	O	V	O	O					
Technology	V	O	V	V	O						
Labour relations	O	O	O	V							
Cybersecurity	V	V	A								
Environmental regulations	O	O									
Project management	A										
Equipment failure											

Table 9.5. Reachability matrix of dimensions

		Dimensions											
	Dimensions	1	2	3	4	5	6	7	8	9	10	11	12
1	Offshore outsourcing risk	1	0	0	1	0	0	0	0	0	0	0	0
2	Quality	0	1	0	0	0	0	0	0	0	0	0	0
3	Inaccurate demand forecast	1	1	1	1	0	0	0	0	1	0	0	0
4	Inventory risk	0	0	0	1	0	0	0	0	0	0	0	0
5	Financial risk	0	0	0	0	1	0	0	0	1	0	1	0
6	Transportation risk	1	0	0	0	0	1	0	0	1	0	1	1
7	Technology	1	1	0	0	0	0	1	0	1	1	0	1
8	Labour relations	0	1	0	1	0	0	0	1	1	0	0	0
9	Cybersecurity	0	1	0	1	0	0	0	0	1	0	1	1
10	Environmental regulations	1	1	0	0	0	0	0	0	1	1	0	0
11	Project management	0	1	1	0	0	0	0	0	0	0	1	0
12	Equipment failure	0	0	0	0	0	1	0	0	0	0	1	1

Table 9.6 Final matrix of reachability with driving dependence powers

	Dimensions											Driving power
Dimensions	1	2	3	4	5	6	7	8	9	10	11	12	
Offshore outsourcing risk	1	0	0	1	0	0	0	0	0	0	0	0	2
Quality	0	1	0	0	0	0	0	0	0	0	0	0	1
Inaccurate demand forecast	1	1	1	1	0	1*	0	0	1	0	1*	1*	8
Inventory risk	0	0	0	1	0	0	0	0	0	0	0	0	1
Financial risk	1*	1*	1*	1*	1	1*	0	0	1	0	1	1*	9
Transportation risk	1	1*	1*	1*	0	1	0	0	1	0	1	1	8
Technology	1	1	1*	1*	0	1*	1	0	1	1	1*	1	10
Labour relations	1*	1	1*	1	0	1*	0	1	1	0	1*	1*	9
Cybersecurity	1*	1	1*	1	0	1*	0	0	1	0	1	1	8
Environmental regulations	1	1	1*	1*	0	1*	0	0	1	1	1*	1	9
Project management	1*	1	1	1*	0	1*	0	0	1*	0	1	1*	8
Equipment failure	1*	1*	1*	1*	0	1	0	0	1*	0	1	1	8
Dependence power	10	10	9	11	1	9	1	1	9	2	9	9	

Furthermore, using stage II of the proposed framework, a list of potential failure modes in automotive supply chain network have been identified and ISM is used to model the relationship among various failure modes of the automotive supply chain.

Table 9.7 Iterations

Criteria	Reachability	Antecedent	Intersection	Level
Iteration I				
1	1,4	1,3,5,6,7,8,9,10,11,12	1	
2	2	2,3,5,6,7,8,9,10,11,12	2	I
3	1,2,3,4,6,9,11,12	3,5,6,7,8,9,10,11,12	3,6,9,11,12	
4	4	1,3,4,5,6,7,8,9,10,11,12	4	I
5	1,2,3,4,5,6,9,11,12	5	5	
6	1,2,3,4,6,9,11,12	3,5,6,7,8,9,10,11,12	3,6,9,11,12	
7	1,2,3,4,6,7,9,10,11,12	7	7	
8	1,2,3,4,6,8,9,11,12	8	8	
9	1,2,3,4,6,9,11,12	3,5,6,7,8,9,10,11,12	3,6,9,11,12	
10	1,2,3,4,6,9,10,11,12	7,10	10	
11	1,2,3,4,6,9,11,12	3,5,6,7,8,9,10,11,12	3,6,9,11,12	
12	1,2,3,4,6,9,11,12	3,5,6,7,8,9,10,11,12	3,6,9,11,12	
Iteration II				
1	1	1,3,5,6,7,8,9,10,11,12	1	II
3	1,3,6,9,11,12	3,5,6,7,8,9,10,11,12	3,6,9,11,12	
5	1,3,5,6,9,11,12	5	5	
6	1,3,6,9,11,12	3,5,6,7,8,9,10,11,12	3,6,9,11,12	
7	1,3,6,7,9,10,11,12	7	7	
8	1,3,6,8,9,11,12	8	8	
9	1,3,6,9,11,12	3,5,6,7,8,9,10,11,12	3,6,9,11,12	
10	1,3,6,9,10,11,12	7,10	10	
11	1,3,6,9,11,12	3,5,6,7,8,9,10,11,12	3,6,9,11,12	
12	1,3,6,9,11,12	3,5,6,7,8,9,10,11,12	3,6,9,11,12	
Iteration III				
3	3,6,9,11,12	3,5,6,7,8,9,10,11,12	3,6,9,11,12	III
5	3,5,6,9,11,12	5	5	
6	3,6,9,11,12	3,5,6,7,8,9,10,11,12	3,6,9,11,12	III
7	3,6,7,9,10,11,12	7	7	
8	3,6,8,9,11,12	8	8	
9	3,6,9,11,12	3,5,6,7,8,9,10,11,12	3,6,9,11,12	III
10	3,6,9,10,11,12	7,10	10	
11	3,6,9,11,12	3,5,6,7,8,9,10,11,12	3,6,9,11,12	III
12	3,6,9,11,12	3,5,6,7,8,9,10,11,12	3,6,9,11,12	III
Iteration IV				
5	5	5	5	IV
7	7,10	7	7	
8	8	8	8	IV
10	10	7,10	10	IV
Iteration V				
7	7	7	7	V

9.5 HIERARCHICAL STRUCTURAL MODEL FOR AUTOMOTIVE RISK DIMENSIONS

The model has been developed by following various steps as shown in Figure 9.1 and discussed in Section 9.3.

9.5.1 Structural self-interaction (SSI) matrix

The relative association among the automotive dimensions is represented as "lead to", which states how one dimension leads to another dimension. Of the 24 potential failure modes, 12 dimensions are selected based on consultations with a total of 6 experts, out of which 4 were from the outsourcing companies and 2 were from academic units. On the basis of responses, SSI matrix is made as presented in Table 9.4. The matrix was formed using V, A, X and O symbols.

9.5.2 Formation of the initial reachability matrix (IRM)

The SSI matrix is converted into (0, 1) matrix, known as the IRM. The symbols representing the relationship among the dimensions in Table 5.3 are substituted by 0 and 1 by following the rules:

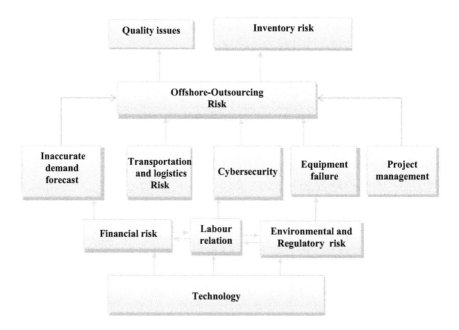

Figure 9.3 ISM model with automotive risk dimensions.

i. If (i, j) in the SSI matrix is represented by V, then (i, j) in the RM will be 1 and (j, i) value is 0.

ii. If (i, j) in the SSI matrix is represented by A, then (i, j) in the RM will be 0 and (j, i) value is 1.

iii. If (i, j) in the SSI matrix is represented by X, then both (i, j) and (j, i) values in the RM will be 1.

iv. If (i, j) in the SSI matrix is represented by O, both (i, j) and (j, i) entries in the RM is 0.

The IRM is shown in Table 9.5. The transitive relation between the dimensions is eliminated and the reachability matrix is obtained in the final from the initial one. Table 8.6 shows the entries in the matrix with the driver and dependence scores of various dimensions. The last column presents the values of driving power and the last row presents the values of dependence power of variables.

9.5.2.1 Partitioning of level

After creating the final reachability matrix, further processing is done to make the hierarchical model based on association among the criteria. To do this, the sets of reachability and antecedent for each dimension are obtained for the identified dimensions and associated levels. Iterations are continued and dimensions where both reachability and intersection sets are the same are positioned at the topmost place in the ISM model. The steps are repeated until all obtained levels of the structural model are determined.

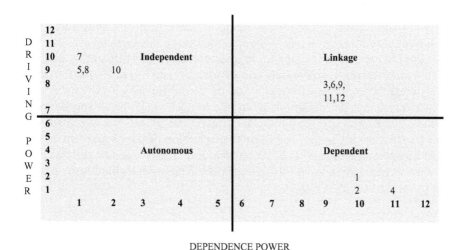

Figure 9.4 MICMAC diagram for dimensions.

In this case, the level identification is completed using 5 iterations for the 12 dimensions. Table 9.7 shows the iterative steps.

9.5.3 Conical matrix

The conical matrix is formulated by joining together risk dimensions which are at a similar level in corresponding rows and columns of the final reachability matrix (FRM).

9.5.4 Diagraph and ISM model development

An intrepretive model is made after developing the conical matrix. Initially, the digraph is made from the conical matrix by eliminating transitivity links, and finally the digraph is obtained. The ISM model as developed from a digraph representing the hierarchy of dimensions is presented in Figure 9.3. Dimension at Level I is positioned at the topmost position in the model followed by dimensions on second position at Level II, dimensions on third position at Level III and similarly the variables are placed up to the last level one by one. In the figure, the dimension quality and inventory risk is placed at the top of the ISM model. Because of the Covid-19 pandemic, companies face shortages of raw material and semiconductor inventories. The shortage in supplies of semiconductors has hampered the automotive companies in a big way and is the major cause for delay in deliveries. The dimension offshore outsourcing is placed at Level II which shows the agreements or contracts with parties overseas were not committed timely because of the spread of the pandemic in different countries across the globe. At Level III, there were five dimensions, viz. cybersecurity, transportation, inaccurate demand forecast, project management and equipment failures. The risk of data breach in IT outsourcing services has resulted in cybersecurity and companies usually find it difficult to manage these risks owing to increased competitive pressure. Transportation also gets badly affected in different parts of the world which affect the supply of raw material and goods from one part of the globe to another part of the globe. The project management activities within the organizations have suffered a lot as they fail to meet production demands and hence have lost businesses. Financial risk, labour relations and environmental risks are shown at Level IV in the model. Finally, the dimension technology is placed at Level V in the ISM model.

9.6 MODEL ANALYSIS AND VALIDATION

To assess the power of dimensions, viz. driver and dependent power and to further classify them under different clusters, namely, autonomous, linkage, dependent and independent, the MICMAC analysis is used. Figure 9.4

presents the results of the MICMAC analysis which is used to validate the ISM model for vendor selection dimensions.

The four clusters with definitions are discussed as under:

1. *Autonomous dimensions*: Those dimensions possessing the weakest powers of driving and dependence fall in the group of autonomous dimensions.
2. *Linkage dimensions*: The dimensions having the strongest powers of driving and dependence falls in the group of linkage dimensions.
3. *Dependent risk dimensions*: This category contains criteria which possess weak driving and strong dependence powers.
4. *Independent risk dimensions*: Here in this category, dimensions possess strong power of driving and weak power of dependence.

   - *Quadrant 1*: No automotive risk dimension is found in this quadrant which states that all the criteria are important. Firms should consider all the risk dimensions for their automotive supply chains.
   - *Quadrant 2*: Among 12 dimensions in the study, 3 dimensions, viz. quality, "inventory" and offshore outsourcing, fall under this quadrant. All three criteria are with less power of driving and high power of dependence. They are placed at the topmost position in the ISM model.
   - *Quadrant 3*: This category consists of five criteria, viz. cybersecurity, transportation, inaccurate demand forecast, project management and equipment failures. All five criteria are affected by low-level criteria in the model and hence they are called linkage criteria.
   - *Quadrant 4*: *Independent category*: In this category, criteria, viz. financial risk, labour relations, environmental and regulatory risk relations and technology are found based upon the results of the ISM model.

9.7 CONCLUSION

The automotive sector has also been hit hard by the Covid-19 pandemic as its supply chain was widely disrupted. Over the years, the automotive supply chain has transformed into a complex multi-national distribution network. The supply chains of automotive firms are multifaceted because a significant number of components, parts to be used in an automobile, are to be sourced from suppliers placed at different tiers, which calls for the coordination of various flows across the supply chain. The collaboration among various supply chain entities for carrying out functions related to procurement, design, production, sales and logistics is essential to reduce the supply chain risk and disruptions. Also, the presence of uncertainties along with natural and man-made disasters have continually disturbed and interrupted the supply chains of automobile manufacturers.

In this chapter, FMEA and ISM approaches are used to identify the dimensions that affect the automotive supply chain considering the Covid-19 pandemic. The identified risks are assessed for their occurrence, severity and detection using the FMEA. The FMEA results portray that in the vendor category, vendor quality has high weighted risk priority number 5.75 followed by economic position weighted risk priority number 4.72. Under logistics and material handling cluster, transportation risk has high weighted risk priority number, i.e., 7.64 followed by labour relations with weighted risk priority number, i.e., 5.71. As during the Covid-19 pandemic, the problems with transportation modes and labour shortages (both skilled and unskilled manpower) were experienced by the companies which result in these risks with high WRPN scores. Under the manufacturer cluster, the dimension offshore outsourcing risk has high weighted risk priority number 7.54 followed by financial risk weighted risk priority number 7.06.

Under the customer cluster, of the three the failure modes (viz. order fulfilment, customer complaints and cybersecurity), cybersecurity has high weighted risk priority number, i.e., 6.13 followed by customer complaints weighted risk priority number, i.e., 5.58. In a digitally connected world, data security has become a major challenge in the adoption of technology among the business houses because of the threats related to data breaches, cyberattacks and malware. To counter these threats, the demand for cybersecurity solutions has increased. According to fortune business insights, the global cybersecurity market is expected to reach $376.32 billion by the year 2029.

In stage II, the FMEA forms the basis for selecting the 12 key risk dimensions to be used further and build an interpretive structural model to study the structural association among them. The outcome of the ISM approach shows that financial risk, environmental risk, labour relations and technology possess high driving power and less dependence so supply chain managers should give more attention towards them. Also, quality, inventory and offshore-outsourcings exhibit strong dependent power and thus they are to be considered as the most important risks and managers should consider them for reducing the risks and mitigating their undesirable effects on automotive supply chains.

FURTHER READING

Cha, S.Y. (2022). The Art of Cyber Security in the Age of the Digital Supply Chain: Detecting and Defending Against Vulnerabilities in Your Supply Chain. In: MacCarthy, B.L., Ivanov, D. (eds) *The Digital Supply Chain* (pp. 215–233). Elsevier, ISBN 9780323916141.

Chen, P.-S., Wu, M.-T. (2013). A modified failure mode and effects analysis method for supplier selection problems in the supply chain risk environment: a case study. *Computers & Industrial Engineering*, 66(4), 634–642.

Dhone, N.C., Kamble, S.S. (2015). Scale development for supply chain operational performance model in Indian automobile industry by using exploratory factor analysis. *International Journal of Logistics Systems and Management*, 22(2), 210–249.

Duhamel, F., Carbone, V., Moatti, V. (2016). The impact of internal and external collaboration on the performance of supply chain risk management. *International Journal of Logistics Systems and Management*, 23(4), 534–557.

Errico, M., De Noni, I., Teodori, C. (2022). SMEs' financial risks in supply chain trade with large companies: the case of Italian automotive component industry. *Journal of General Management*, 47(2), 126–137.

Ghadge, A., Weiß, M., Caldwell, N.D., Wilding, R. (2019). Managing cyber risk in supply chains: a review and research agenda. *Supply Chain Management: An International Journal*, 25(2), 223–240. https://doi.org/10.1108/scm-10-2018-0357.

Kumar Sharma, S., Bhat, A. (2014). Supply chain risk management dimensions in Indian automobile industry: a cluster analysis approach. *Benchmarking: An International Journal*, 21(6), 1023–1040. https://doi.org/10.1108/BIJ-02-2013-0023.

Luthra, S., Luthra, S., Haleem, A. (2015). Hurdles in implementing sustainable supply chain management: an analysis of Indian automobile sector. *Procedia - Social and Behavioral Sciences*, 189, 175–183.

Pandey, A., Sharma, R.K. (2017). FMEA-based interpretive structural modelling approach to model automotive supply chain risk. *International Journal of Logistics Systems and Management*, 27(4), 395–419.

Pandey, S., Singh, R.K., Gunasekaran, A., Kaushik, A. (2020). Cyber security risks in globalized supply chains: conceptual framework. *Journal of Global Operations and Strategic Sourcing*, 13(1), 103–128. https://doi.org/10.1108/JGOSS-05-2019-0042.

Rosangela, M.V., Lucato, W.C., Ganga, G.M.D., Alves Filho, A.G. (2020). Risk management in the automotive supply chain: an exploratory study in Brazil. *International Journal of Production Research*, 58(3), 783–799. https://doi.org/10.1080/00207543.2019.1600762.

Salehi Heidari, S., Khanbabaei, M., Sabzehparvar, M. (2018). A model for supply chain risk management in the automotive industry using fuzzy analytic hierarchy process and fuzzy TOPSIS. *Benchmarking: An International Journal*, 25(9), 3831–3857. https://doi.org/10.1108/bij-11-2016-0167.

Sawik, T. (2013). Selection and protection of suppliers in a supply chain with disruption risks. *International Journal of Logistics Systems and Management*, 15(January), 143–159.

Sharma, R.K. (2022). Examining interaction among supplier selection strategies in an outsourcing environment using ISM and fuzzy logic approach. *International Journal of System Assurance Engineering*, 13, 2175–2194. https://doi.org/10.1007/s13198-022-01624-2.

Surange, V.G., Bokade, S.U. (2022). Ranking of Critical Risk Factors in the Indian Automotive Supply Chain Using TOPSIS with Entropy Weighted Criterions. In: Chaurasiya, P.K., Singh, A., Verma, T.N., Rajak, U. (eds) *Technology Innovation in Mechanical Engineering. Lecture Notes in Mechanical Engineering*. Springer, Singapore. https://doi.org/10.1007/978-981-16-7909-4_46.

Wang, L., Foerstl, K., Zimmermann, F. (2017). Supply Chain Risk Management in the Automotive Industry: Cross-Functional and Multi-tier Perspectives. In: Abele, E., Boltze, M., Pfohl, H.C. (eds) *Dynamic and Seamless Integration of Production, Logistics and Traffic*. Springer, Cham. https://doi.org/10.1007/978-3-319-41097-5_7.

Appendix

Table A.1 Detail of stakeholders

S. No. of respondent	Specialized area of organization[a]
R1	Information technology and security services
R2	Business analytics and data mining service
R3	R&D and business solutions
R4	Business finance and security services
R5	Information technology service
R6	KPO outsourcing
R7	Information technology and security services
R8	Business development and sales support service
R9	Business process outsourcing
R10	Inventory service management
R11	KPO outsourcing
R12	Business analytics and data mining service
R13	Social media monitoring service
R14	Information services and knowledge management
R15	KPO outsourcing
R16	Business analytics and data mining service
R17	Information technology and security services
R18	Business process outsourcing
R19	R&D and business solutions
R20	Business analytics and data mining service
R21	Business finance and security services
R22	Business analytics and data mining service
R23	KPO outsourcing
R24	Information technology and security services
R25	R&D and business solutions
R26	Business analytics and data mining service
R27	R&D and business solutions
R28	Information technology and security services

[a] For protecting the privacy of respondents, the identities of respondent organizations are concealed.

Table A.2 Summary of selected views regarding political risk

S. No.	Summarized views	Stakeholders
1	The countries affected by terrorism are not considered for the global business for offshoring outsources projects. International terrorist movement reduces the collaborative work due to feeling of insecurity.	[R1], [R4], [R10], [R17], [R22], [R24], [R25], [R27]
2	Because of a civil war, unexpectedly all the offshore activities may come to a halt, and it results in total failure of outsourcing projects.	[R3], [R3], [R5], [R16] [R10], [R11], [R14], [R18], [R22]
3	In some countries, even for legalized work, money is supposed to be paid to sanctioned authority due to their corrupt behaviour.	[R7], [R10], [R18], [R19], [R20], [R23], [R24], [R27]
4	It is always possible that data/process of a client organization may be leaked by the employees of the service provider organization and it reduces the industrial relationship.	[R2], [R3], [R5], [R9], [R14], [R16], [R27], [R28]
5	It happens many times that in developing countries their value of currency abruptly changes and it affects the export–import of the asset specificity.	[R3], [R7], [R13], [R17], [R21], [R22], [R24].
6	Differences between the countries have its direct impact on collaborative projects.	[R3], [R10], [R17], [R18], [R12], [R14], [R28].
7	Favouritism policies towards domestic organizations provide a rough environment for the client organization.	[R5], [R6], [R10], [R12], [R14], [R16].
8	The main motive of offshore outsourcing is to exploit the best available human resource from anywhere. Many times, due to many reasons, it is not workable.	[R1], [R5], [R8], [R12], [R22], [R23],[R26]
9	Many times, it happens that the relation between the countries is disturbed due to some reason and it creates problems for offshore outsourcing task.	[R2], [R10], [R11], [R17], [R21], [R28].

(Continued)

Table A.2 (Continued) Summary of selected views regarding political risk

S. No.	Summarized views	Stakeholders
10	Uncertainty in political environment reduces the prospects of offshore outsourcing projects.	[R1], [R7], [R9], [R10], [R15], [R20] [R24], [R25], [R27].
11	The hostile relationship between the two countries has a direct impact on the collaboration assignment.	[R2], [R3], [R5], [R7], [R20], [R25], [R26], [R27]
12	Because of demographic changes occurring in the country of the service provider, chances to continue further operations becomes bleak.	[R1], [R4], [R5], [R16], [R22], [R24], [R25], [R28]
13	Many times, there is a compulsion from the country of the service provider organization to start the activity from a remote location/underdeveloped areas and it reduces the environment of business scenario.	[R2], [R3], [R5], [R9], [R11], [R14], [R17].
14	Because of corruption, transparency about the decision-making diminishes and it creates frustration among stakeholders and loses business.	[R3], [R4], [R10], [R15], [R17], [R20], [R22], [R28]
15	Adverse monetary policies may affect the viability of projects.	[R3], [R5], [R6], [R9], [R13], [R15], [R18], [R22].
16	Many times, work of outsourcing is not welcomed by employees of the client organization, and it further deteriorates relation between employees and their employer.	[R3], [R5], [R7], [R8], [R12], [R16], [R18], [R27].

Note: Numbers in brackets [ ] represent the serial numbers of respondents from Table A.1.

Table A.3 Summary of selected views regarding risk due to cultural differences

S. No.	Summarized views	Stakeholders
1	Communication problems, for instance, in offshoring business from developing countries may lead to the development of vendor–vendee relationship risk.	[R3], [R5], [R11], [R15], [R17], [R19], [R20], [R28].
2	High turnover of the employees results because of working hours leading to odd timings across the globe.	[R4], [R10], [R12], [R15], [R21], [R23], [R28].
3	The nature of communicating language, i.e., verbal or non-verbal, differentiates one team from another team and the connotation given to gestures; differ obviously from one culture to another culture.	[R3], [R4], [R6], [R10], [R15], [R22].
4	Many times, balance between sense of self and the work is disturbed due to ego clash and other factors between the employees of different organizations.	[R1],[R4], [R7], [R8], [R11]. [R24],[R25],[R27]
5	Many times, food and eating habits make distance among employees of both organizations.	[R3], [R6], [R7], [R13], [R19], [R27].
6	The food habits of teams often differ among cultures as evident from the way in which they prepare, present and eat food of their choice.	[R3], [R4], [R6], [R7], [R8], [R12], [R15], [R17], [R24].
7	Even while speaking the same language, misunderstanding may develop due to different cultural assumptions and jargon terminology.	[R1], [R3], [R10], [R8], [R10] [R13], [R14], [R16],[R25],[R27]
8	Belief also exerts a very strong influence on the wrong behaviour between the two cultures.	[R18], [R19], [R20], [R22], [R24], [R25].

(Continued)

Table A.3 (Continued) Summary of selected views regarding risk due to cultural differences

S. No.	Summarized views	Stakeholders
9.	Due to different communication languages and cultures, members of one group may make distance from members of the other groups.	[R3], [R5], [R7], [R8], [R12], [R14], [R20].
10.	Many times, strict dress code belonging to one culture gives uncomfortable feeling towards the employee of the other culture. Unacceptable dress and appearances also increase the gap between experts belonging to different cultures.	[R1], [R3], [R6], [R10], [R12], [R14], [R16], [R21], [R27]
11.	During discussions, complex situations might be not addressed properly due to language problem.	[R9], [R12], [R15], [R16], [R24], [R25], [R27]
12.	Cultures interpret time and time consciousness differently, and it creates a wide gap between the organizations.	[R2], [R8], [R10], [R13], [R15], [R22], [R24].
13.	People of all cultures have a strong belief and attitude for supernatural power which is apparent from the religious practices followed by them.	[R2], [R3], [R7], [R9], [R10], [R13], [R19], [R27]
14.	In the collaborative work, the team leader shows undue favour towards their countries' fellows, and it creates dissatisfaction among other members.	[R2], [R3], [R4], [R6], [R14], [R15], [R18], [R20], [R28].

Note: Numbers in brackets [ ] represent the serial numbers of respondents from Table A.1.

Table A4 Summary of selected views regarding opportunistic behaviour risk

S. No.	Summarized views	Stakeholders
1	Service provider knows about the switching cost of the client organization, and it enhances their opportunistic behaviour.	[R3], [R4], [R6], R10], [R16], [R18], [R19], [R20], [R21], [R23].
2	Inadequate training to handle offshore outsourcing projects induces a higher degree of opportunistic risk to the client organizations.	[R1], [R4], [R7], [R8], [R12], [R14], [R15].
3	When span of control with respect to project execution is more lop-sided, it disrupts the functioning of the service provider organization.	[R5], [R8], [R10], [R11], [R14], [R20], [R22].
4	Everything cannot be put on paper so there is huge freedom to the service provider if they want about decision-making.	[R2], [R3], [R8], [R9], [R11], [R14], [R24].
5	Travel time is a big obstacle to frequently meet the people of the service provider organization.	[R4], [R5], [R10], [R16], [R22], [R23].
6	Whenever there is turnover of employees at the service provider organization, communication gap pinches a lot.	[R3], [R10], [R12], [R18], [R21], [R23], [R25].
7	There are chances that the service provider may join one of the competitors of the client organization after completion of the contract.	[R1], [R5], [R10], [R15], [R17], [R18], [R20], [R21].
8	Overburdening the employee of the service provider organization with multiple tasks also deteriorates the output of the assigned task.	[R2], [R7], [R8], [R13], [R19], [R20], [R22], [R23].
9	Due to switching costs, after sometime, service providers do not execute things properly.	[R1], [R5], [R7], [R12], [R13], [R22], [R23].

(Continued)

Table A4 (Continued) Summary of selected views regarding opportunistic behaviour risk

S. No.	Summarized views	Stakeholders
10	Incomplete or partial work specifications result when both the organizations are in conflict of goals.	[R1], [R4], [R10], [R17], [R19], [R20], [R22], [R24], [R25].
11	There is a risk of inadequate funding or investment by the service provider to the client organization, not getting what was required as per contract obligations and thus providing the limited benefits.	[R3], [R6], [R12], [R15], [R17], [R20], [R22], [R23].
12	Incomplete work specifications give a free hand to take advantage of fulfilling its own objective to the service provider and it may differ from benefits of the client organization.	[R2], [R6], [R7], [R13], [R19], [R25].
13	Many times, it has been observed that the service provider organization takes advantage of the communication gap.	[R1], [R2], [R5], [R10], [R15], [R24],[R25],[R27]
14	Complexity of a task increases confusion and leads to the free hand of the service provider to take advantage of this.	[R1], [R4], [R7], [R9], [R10], [R12], [R15], [R21], [R23], [R25].
15	Client organization loses control on processes and methods to the service provider due to geographical distances and it looks as a worse decision by the client organization.	[R2], [R5], [R7], [R10], [R15], [R17], [R18], [R20], [R22], [R27].

Note: Numbers in brackets [ ] represent the serial numbers of respondents from Table A.1.

Table A.5 Summary of selected views regarding intellectual property risk

S. No.	Summarized views	Stakeholder
1	Shifting research and development process and laying off of their employees, client organization loses their assets related to intellectual property.	[R1], [R8],[R14], [R11], [R14], [R22], [R26].
2	Owing to movement of persons from one place to another place, the chances of breach in contract may be possible.	[R5], [R10], [R15], [R19], [R20], [R22], [R23], [R25].
3	Sometimes in litigation process, when the behaviour of service provider organization's country shows "Particularism" and takes unethical favour of the service provider, it creates a problem for the client organization.	[R2], [R9], [R10], [R15], [R22], [R23], [R27], [R28].
4	By acquiring the knowledge of the task, the service provider organization rises as a competitor of the client organization.	[R1], [R10], [R12], [R13], [R16], [R24], [R25], [R27]
5	Breach of IPR rights, by the service provider organization due to opportunistic behaviour, is a big risk for client organization.	[R6], [R12], [R15], [R20], [R22], [R23].
6	Easy availability of state-of-the-art high-end information technology aggravates chances of the risk of IPR.	[R4], [R13],[R15], [R16], [R24],[R26]
7	With new hacking methods, information related to intellectual property may leak at any point for organization operating globally.	[R5], [R10], [R14], [R22], [R28].

(Continued)

Table A.5 (Continued) Summary of selected views regarding intellectual property risk

S. No.	Summarized views	Stakeholder
8	In knowledge-oriented work, on many occasions, threats are created by employee of the service provider regarding intellectual property.	[R3], [R7], [R16], [R18], [R21], [R23], [R27].
9	Whenever there is complex and huge dimension of the assigned tasks, chances of IP risk increase by many folds.	[R1], [R5], [R7], [R9], [R11], [R12], [R23].
10	The service provider of offshore outsourcing may emerge as competitor of the client organization by using its technology.	[R2], [R4], [R6], [R9], [R10], [R15], [R19], [R20], [R23].
11	There is a chance that the confidential data or technology may be passed on to competitors of the client organization by the service provider.	[R1], [R4], [R8], [R12], [R15], [R18], [R21], [R24].
12	Whenever there is an involvement of many people, there is a chance of leakage of confidential information which may be the sole property of the client organization.	[R1], [R3], [R5], [R6], [R10], [R14], [R17], [R26].
13	By transferring R&D related work, client organizations lose their upper hand on the designing section.	[R2], [R8], [R10], [R15], [R17], [R28].
14	Offshore outsourcing starts from shifting low level of work to the service provider organization, and later on, further improvement can only be done by the service provider which is a huge loss at strategic level.	[R2], [R4], [R9], [R10], [R12], [R13], [R15], [R27]

Note: Numbers in brackets [ ] represent the serial numbers of respondents from Table A.1.

Table A.6 Summary of selected views regarding financial risk

S. No.	Summarized views	Stakeholders
1	Whenever there is a high-end work or change of technical specifications of the work, it causes high adaptation cost for the client organization.	[R2], [R4], [R6], [R8], [R10], [R12], [R14], [R20], R25].
2	Transition and management cost, especially at the starting stage, may be beyond expectations, which causes financial risk.	[R5], [R7], [R11], [R14], [R24].
3	Due to costing of asset specificity, switching cost increases greatly and the service provider organization may take advantage because of this reason.	[R3], [R10], [R13], [R17], [R21], [R22], [R24].
4	Because the service provider organization belongs to a different country, any type of communication system may be a costly affair.	[R8], [R11], [R15], [R18], [R20], [R27].
5	Many times, at the beginning, the total calculation about work is not done correctly and afterwards, when things change, it creates problem.	[R3], [R7], [R8], [R13], [R20], [R25],[R28]
6	Each and every minute changes cannot be put on a formal contract; the service provider organization takes advantage of this, and it may be a loss for the client organization.	[R3], [R9], [R10], [R15], [R19], [R23], [R27].
7	There are conflicts about standards of performance among both organizations which may create a misunderstanding on many occasions.	[R7], [R12], [R14], [R18], [R19], [R20], [R28].
8	Trade-off between the organization cultures of both organizations takes time, and it reflects as transition and management cost.	[R2], [R10], [R12], [R17]. [R22],[R26],[R27]
9	Financial metrics across firms differ, and it converts into a huge invisible cost for the client organization.	[R4], [R11], [R13], [R15], [R19], [R21], [R27].
10	Whenever there are disputes and litigations due to any reasons, these place a financial burden on the client organization.	[R11], [R14], [R16], [R18], [R19], [R28].
11	Adaptation and switching costs become high due to change of technical specifications of the work solutions resulting in financial burden.	[R2], [R4], [R10], [R16], [R19], [R22], [R28].
12	After completion of the contract, much confidential information is kept by the service provider organization, and due to this, it might increase the next contract amount.	[R5], [R12], [R15], [R16], [R20], [R27].

Note: Numbers in brackets [ ] represent the serial numbers of respondents from Table A.1.

Table A.7 Summary of selected views regarding organization structural risk

S. No.	Summarized views	Stakeholder
1	On many occasions, service provider's accepted work is beyond their capacity, and further, it deteriorates the quality of assigned task.	[R2], [R5], [R10], [R14], [R20],[R23],[R27]
2	Many times, geographic distance is a hurdle for resolving regulatory issues between both organizations which belong to different countries.	[R1], [R5], [R10], [R17], [R22], [R26].
3	Inconsistency in assignment of tasks among individuals inhibits the development of infrastructure required to complete the task, and it delays the completion of project.	[R4], [R7], [R8], [R10], [R14], [R17], [R20], [R25].
4	It may happen that the service provider takes up many assignments simultaneously and does not provide proper resources to each task.	[R5], [R9], [R13], [R14], [R19],[R21],[R27]
5	Mismatch between organizations' work culture creates issues of conflict, and due to this, assigned task suffers.	[R3], [R4], [R7], [R8]. [R20],[R25],[R27]
6	Lack of experience was shown by many employees of the service provider organization because the assigned work was having new terminology and new environment for the task.	[R1], [R5], [R10], [R13], [R22], [R26].
7	Incompetency of service providers to fulfil the task may be a very costly affair for the client organization.	[R5], [R8], [R10], [R11], [R12], [R20], [R22].
8	Reduction in quantity as well as quality of human resource by the service provider deteriorates the results of offshore outsourcing by the service providers.	[R8], [R12], [R14], [R17], [R19], [R20], [R27].
9	Turnover of employee of the client organization and not employing the competent person in place of them result in deterioration of the output of the assigned project	[R1], [R4], [R7], [R9], [R12], [R14], [R15],[R19]
10	Differences in perceptions about decision-making among organizations affect the quantity and quality of resources for a task, and it creates conflict.	[R1],[R3], [R8], [R12], [R15], [R20], [R21], [R28].
11	The lack of coordination among various entities in an organization results in organizational structural risk.	[R1], [R4], [R7], [R10], [R12], [R21], [R22],[R27]
12	Most service providers do not take into consideration all the available possibilities and consider only organizational resources that are not appropriate to accomplish their goals.	[R5], [R11], [R17], [R19]. [R21],[R25],[R27]
13	Difference in organization structures of client and service provider organization creates mismatch between their working styles and results in incompatibility issues.	[R1], [R7], [R9], [R15], [R22], [R28].

Note: Numbers in brackets [ ] represent the serial numbers of respondents from Table A.1.

Table A 8 Summary of selected views regarding operational risk

S. No.	Summarized views	Stakeholders
1	Lack of experience shown by employees of the service provider organization when assigned work included new terminologies and new environment for the task results in operational risk.	[R3], [R5], [R7], [R9], [R10],[R14], [R20],[R25],[R28]
2	When the tasks are not assigned with proper priorities, there are chances of delay in the delivery.	[R1], [R5], [R7], [R9], [R11], [R13], [R12], [R14], [R19].
3	The difference in expected improvement in production quality with cost reduction and service-level improvement between service provider and client organization may also give rise to conflict which results in low performance.	[R3], [R5], [R7], [R12], [R15], [R16], [R22], [R24], [R26],[R24]
4	Many times, non-transparency in the process by service provider generates operational risk because it is almost a black-box approach system.	[R2], [R3], [R11], [R12], [R15], [R22], [R25].
5	Frequent turnover of the employee generates discontinuity in the levels of communication, so there is a negative effect on the operation.	[R1], [R10], [R12], [R15], [R17], [R20], [R28].
6	Incompetency of service providers to fulfil the task may be a very costly affair for the client organization.	[R2], [R6], [R14], [R17], [R24], [R25],[R28]
7	According to the agreement, if the expertise of the employee of the service provider is not consistent with time, it reduces the quality of work which results in operational risk.	[R3], [R5], [R7], [R8], [R12], [R14], [R21].
8	Mostly in offshore outsourcing, work is done across different time zones and it disturbs personal time of any one group, due to which output is not up to the mark.	[R1], [R8], [R13], [R17], [R21], [R22], [R28].

(Continued)

Table A 8 (Continued) Summary of selected views regarding operational risk

S. No.	Summarized views	Stakeholders
9	Many times, operational facilities are not up to the mark and the result is deteriorated output.	[R7], [R12], [R15], [R17], [R19], [R20],[R24].
10	High turnover rate in an organization generates discontinuity in the level of communication, so there is negative effect on the operation.	[R4], [R10], [R14], [R16], [R19],[R20], [R28]
11	Operational risk results in suboptimal output which is obtained as a result of complex operations and geographical distance between the vendor and client.	[R3], [R10], [R17], [R19], [R22], [R24], [R27].
12	If time zone difference is greater, then the adverse effect on client quality of service is also greater.	[R2], [R6], [R10], [R17], [R19],[R24],[R28]
13	Poor delivery performance, lack of competency to fulfil task and process fragmentation result in operational risk.	[R1], [R7], [R7], [R9], [R12], [R14], [R27].
14	Numerous times, this risk comes into account when client loses control over operations due to offshore outsourcing.	[R6], [R7], [R10], [R15], [R21],[R24],[R28]
15	Lack of competency to fulfil a task can be considered as not having the required characteristics for performing a given task, activity or role successfully.	[R3], [R5], [R7], [R12], [R22], [R25], [R27],[R28].

Note: Numbers in brackets [ ] represent the serial numbers of respondents from Table A.1.

Table A.9 Summary of selected views regarding compliance and regulatory risk

S. No.	Summarized views	Stakeholder
1	Incompatibility between client and vendor countries about rules and regulations gives advantage to the service provider.	[R4], [R7], [R10], [R12], [R14], [R16].
2	In case if there is no link between service provider and investors, then only the client organization is liable for the wrong doings of the service provider.	[R1], [R6], [R13], [R17], [R21], [R22], [R28].
3	Due to the global nature of offshoring projects, there is a chance of fraud and breach in security of confidential matters at any stage of the project.	[R2], [R12], [R14], [R17], [R24], [R25], [R27]
4	Worsening law-and-order conditions in the country of the service hampers the corrective steps taken by the client organization.	[R2], [R6], [R8], [R9], [R10], [R11], [R18].
5	During the project execution, it may happen that tax structure of the service provider's country changes and extra cost is incurred in the project cost.	[R2], [R7], [R12], [R14], [R27], [R28]
6	Data are often shared and given to the service provider for advanced research, and there is a risk of confidentiality, which remains permanently.	[R8], [R11], [R15], [R17], [R20], [R22].
7	During offshore outsourcing, in the beginning, certain things might be not still clear with the client, so later on, the client cannot take advantage of rules and regulations.	[R1], [R2], [R7], [R10], [R16], [R18], [R20], [R27].
8	Due to non-compliance of the task up to the prescribed time, the brand reputation of client is at stake.	[R1], [R7], [R10], [R14], [R18], [R19], [R28].
9	Even though rules and regulations exist, authorities do not implement them due to various reasons like favouritism and corruption.	[R1], [R4], [R6], [R9], [R14], [R17], [R22], [R24].

Note: Numbers in brackets [ ] represent the serial numbers of respondents from Table A.1.

Table A.10 Summary of selected views regarding loss of core professionals

S. No.	Summarized views	Stakeholder
1	Due to shifting of technical work at the service provider, reshuffling of the experts of that field of work at other places is a non-utilization of their expertise.	[R1], [R12], [R15], [R19], [R20], [R25].
2	Professionals always enjoy work of their choice, and whenever they are assigned non-challenging task, they are get bored, leading to a decrease in efficiency.	[R2], [R10], [R12], [R15], [R21], [R23], [R28].
3	Lack of communication also leads to loss of core professionals due to misunderstanding about scope in future within the client organization.	[R1], [R4], [R12], [R19], [R20], [R23], [R25].
4	Layoff scenario at the client side is a lose–lose scenario because the client loses their core professionals and employees lose their expertise at work.	[R3], [R7], [R10], [R11] [R15], [R17], [R21]
5	Due to process fragmentation between the client and the service provider organizations, size of work gets reduced from the client's side and many professionals are forced to sit on benches.	[R2], [R6], [R17], [R18], [R19], [R25].
6	Shifting the lower level work to the service provider further ensures that modification and improvement are done only by the service provider at the later stages.	[R5], [R8], [R14], [R15], [R16], [R19], [R23],[R27]
7	Offshore outsourcing reflects do-or-buy phenomenon, and due to this, after sometime, the expertise shifts to the service provider from the client organization.	[R2], [R11], [R20], [R24], [R25], [R25]
8	Loss of knowledge pool because of lay offs and exploitation of experts.	[R1], [R3], [R10], [R14], [R15], [R19], [R27]
9	After a certain period of time, it is mandatory for the client organization to accomplish the task by the service provider, due to layoff of competent experts by the client organization.	[R4], [R7], [R11], [R13], [R16], [R18], [R24]
10	Layoff(s) exercised by the service provider companies or by employer results in loss of core professionals.	[R2], [R4], [R11], [R17] [R19], [R20], [R27]
11	The risk arises with loss of core professionals or knowledge pool the organization has to handle the outsourcing activities.	[R5], [R8], [R10], [R13], [R15], [R21], [R28]
12	Exploitation of experts also leads to loss of core professionals.	[R2], [R7], [R11], [R14], [R15], [R19], [R26]
13	The loss of critical knowledge is seen as the greatest source of workforce-related offshore outsourcing risk.	[R1], [R8], [R11], [R14], [R15], [R21], [R25]

Note: Numbers in brackets [ ] represent the serial numbers of respondents from Table A.1.

Table A.11 Summary of selected views regarding cybersecurity in digital supply chains

S. No.	Summarized views	Stakeholder
1	Information security poses a huge challenge in business as much of our information services are outsourced.	[R3], [R7], [R11], [R20], [R24].
2	Today, firms have become aware of risks associated with cybersecurity, and hence they have raised their budgets to counter the risks.	[R2], [R11], [R14], [R15], [R21], [R22], [R26].
3	Digital supply chains add complexity to business and have become more challenging to manage.	[R9], [R7], [R10], [R11] [R19], [R23], [R27]
4	The growing capability of cloud-based IT systems has advanced the threats related to cybersecurity, thus making cybersecurity risk concerns dominant in all forms of IT outsourcing business.	[R4], [R6], [R7], [R19], [R21], [R25], [R28].
5	Because of issues related to data privacy, compliance and access controls causes obstacle in adoption of cloud services.	[R5], [R8], [R14], [R15], [R17], [R19], [R24], [R27].
6	Lack of trust and transparency concerns on cybersecurity are more for smaller business firms as they outsource more of their IT needs.	[R2], [R7], [R17], [R18], [R24], [R26].
7	Incidents related to hacking are the dominant forms of risks which are behind healthcare industry data breaches, followed by prohibited internal releases.	[R4], [R9], [R17], [R24], [R24], [R27]
8	In a digital supply chain, data security has become a major challenge in the adoption of technology because of the threats related to data breaches, cyberattacks and malware.	[R1], [R8], [R17], [R20], [R24], [R28].
9	Insider threats are caused by workers in which they use their authorized access to harm the organization.	[R4], [R7], [R11], [R13], [R16], [R27].
10	Because of divergent ICT standards between economies related to digitization systems, the cross-country risks pose implications for national security.	[R1], [R4], [R11], [R17] [R19], [R24], [R27]
11	The software vulnerabilities embedded into the software during the design or implementation phase possess considerable risk when some of the systems are neither updated nor fixed at all.	[R3], [R9], [R19], [R21], [R25],[R28]
12	The contextual risks associated with supporting processes such as procurement, staffing, funding, training and development considerably affect the digital supply chains.	[R5], [R8], [R14], [R15], [R17], [R24].
13	Insider threats deliberate or premeditated because of human factors can pose the biggest threat to a firm's cybersecurity.	[R2], [R4], [R7], [R19], [R21], [R23], [R28].

Note: Numbers in brackets [ ] represent the serial numbers of respondents from Table A.1.

Table A.12 Core processes in offshore outsourcing

S. No.	Processes related to offshore outsourcing	S. No.	Processes related to offshore outsourcing
1	Selecting a service provider organization	16	Synergies across services
2	Attrition of the existing workforce	17	Extensive alignment
3	Dealing with cultural differences	18	Decommission of existing systems
4	Managing an offshore contract	19	Product evaluation
5	Selecting the nature of work to be outsourced	20	Contract negotiation
6	Transitioning work at offshore	21	Synchronizing activities among people and work in an organized manner for timely completion
7	Adaptation with offshore outsourcing environment	22	Monitoring the ongoing outsourcing relationship
8	Costing methods due to offshore outsourcing	23	Creating and managing the contract
9	Approval of stakeholders for offshore outsourcing	24	Establishing service-level metrics
10	Transferring of knowledge domain of an assigned task	25	Designing cycle time of work
11	Alignment of governance capabilities	26	Understanding about processes and communication protocols
12	Integration-centric engineering	27	Handling cybersecurity risk
13	Standardization of IT processes and communication protocols	28	Managing forecast errors
14	Organizational redesign	29	Improving quality of services
15	Ability to move to standardized practices	30	Handling insider threats

Table A.13 Outcomes of learning

S. No.	Offshore outsourcing risk	No. of dimensions	Description of dimensions	
1	Intellectual Property risk	6	• Project size • Research and development process	• Technical knowhow • Lawless environment	• Opportunistic behaviour • Information technology
2	Compliance and regulatory risk	6	• Brand reputation • Meeting demand of investors	• Fraud and security breaches • Defence against legal repercussions	• Change in tax structure • Data protection issues
3	Political risk	8	• Civil disturbance • Fiscal and Monetary policy • Terrorist events	• Industrial labour relations • Public policy of the host country • Corrupt behaviour and bribery	• Foreign affairs • Availability of suitable human resource
4	Operational risk	7	• Poor delivery performance • Lack of competency to fulfil task • Process fragmentation	• Conflict of objectives • Staff turnover	• Liabilities and litigations • Poor quality of service
5	Risk due to culture differences	8	• Language and information exchange • Working habits and work ethics • Learning strategies	• Self-esteem • Dress code and professional look • Food habits	• Managing time zone differences • People's values, beliefs and attitudes

(Continued)

Table A.13 (Continued) Outcomes of learning

S. No.	Offshore outsourcing risk	No. of dimensions	Description of dimensions
6	Opportunistic behaviour risk	8	• Communication gap • Geographical distances • Incomplete work specifications • Level of control • Post-contractual behaviour • Potential transaction costs • Underinvestment in project • Switching cost
7	Organization structural risk	6	• Regulatory issues • Incompatibility • Inappropriate resource allocation • Deficiency in capabilities • Delay in completion of a project • Reduction in human capital
8	Financial risk	8	• Switching cost • Transition and management cost • Layoff cost • Service provider selection cost • Measurement problems • Adaptation cost • Costly contractual amendments • Disputes and litigations
9	Loss of core professionals	6	• Reduction in work size • Loss of knowledge pool • Layoff • Lack of communication • Misuse of experts • Non-challenging task
10	Cybersecurity risk	7	• Within-country risk • Software vulnerabilities • Contextual risk • Physical threats • Insider threats because of humans • Cross-country risk. • Cyber assaults or attacks

Table A.14 Action and Performance Table Related to SAP-LAP Analysis

S. No.	Offshore outsourcing risk (a)	Action to mitigate offshore outsourcing risk (b)	Performance (c)
1	Intellectual property risk	Monitor opportunistic behaviour of the service provider organization	Increased technical knowhow Enhanced competitive gains
2	Political risk	Analyse domestic polices of host country Analyse foreign affairs and relations	Improved international relations Improved fiscal and monetary policies Improved bilateral coordination
3	Operational risk	Solve conflict of objectives Improve quality of service	Enhances service quality Enhances delivery performance
4	Risk due to culture differences	Foster belief and attitude Language and information exchange Managing time zone differences	Better work habits and practices Improved people's values, beliefs and attitudes
5	Opportunistic behaviour risk	Reduce communication gap between organizations and their functions	Improved level of control Complete work specification Reduce switching costs
6	Organization structural risk	Appropriate resource allocation among departments/business functions	Facilitates compatible work styles. Facilitates timely completion of projects
7	Financial risk	Sort out disputes and litigations with clients, if any Avoid costly contractual amendments	Reduce transition and management cost Reduce adaptation cost
8	Compliance and regulatory risk	Prepare legal defence against compliance and regulatory risks	Increases brand reputation Meets demand of investors/clients
9	Loss of core professionals	Retain knowledge workers or core professionals	Reduce layoffs! Reduce lack of communication
10	Cybersecurity risk	Use of system modularity and information technology infrastructure	Reduce insider threats Reduce with in country or geopolitics risk

Bibliography

Abdullah, L.M., Verner, J.M. (2012). Analysis and application of an outsourcing risk framework. *Journal of Systems and Software*, 85(8), 1930–1952.

Aggarwal, A. (2011). *Legal Process Outsourcing*. White Paper. Evalueserve Ltd.

Albrecht, D.J. (2018). Designing Organizations: Outsourcing (White Paper). Alonos Corporation, Dallas, TX. Retrieved from: alonos.com/resources [Accessed 22 Nov. 2022].

Ambos, B., Ambos, T.C. (2011). Meeting the challenge of offshoring R&D: an examination of firm- and location-specific factors. *R&D Management*, 41(2), 107–119.

Ancarani, A., Di Mauro, C., Fratocchi, L., Orzes, G., Sartor, M. (2015). Prior to reshoring: a duration analysis of foreign manufacturing ventures. *International Journal of Production Economics*, 169, 141–155.

Aouadni, S., Aouadni, I., Rebai, A. (2019). A systematic review on supplier selection and order allocation problems. *Journal of Industrial Engineering International*, 15, 267–289.

Ayağ, Z., Samanlioglu, F. (2016). An intelligent approach to supplier evaluation in automotive sector. *Journal of Intelligent Manufacturing*, 27, 889–903.

Barney, S., Mohankumar, V., Chatzipetrou, P., Aurum, A., Wohlin, C., Angelis, L. (2014). Software quality across borders: three case studies on company internal alignment. *Information and Software Technology*, 56(1), 20–38.

Bhattacharya, A., Chetty, P. (2019). Order fulfilment strategies in supply chain management. [online] Project Guru. Available at: https://www.project-guru.in/order-fulfilment-strategies-supply-chain-management/ [Accessed 03 Dec. 2022].

Bhattacharya, A., Singh, P.J., Bhakoo, V. (2013). Revisiting the outsourcing debate: two sides of the same story. *Production Planning & Control*, 24(4–5), 399–422.

Bomhard, D., Daum, A. (2021). Cybersecurity in outsourcing and cloud computing: a growing challenge for contract drafting. *International Cybersecurity Law Review*, 2, 161–171.

Bruccoleri, M., Perrone, G., Mazzola, E., Handfield, R. (2019). The magnitude of a product recall: offshore outsourcing vs. captive offshoring effects. *International Journal of Production Research*, 57, 4211–4227. https://doi.org/10.1080/0020 7543.2018.1533652.

Burger W. (2007). Offshoring and outsourcing to INDIA. Global Software Engineering, 2nd IEEE International Conference on Global Software Engineering, pp. 173–176.

Cai, S., Ci, K., Zou, B. (2011). Producer services outsourcing risk control based on outsourcing contract design: industrial engineering perspective. *Systems Engineering Procedia*, 2, 308–315.

Cappelli, P. (2011). HR sourcing decisions and risk management. *Organizational Dynamics*, 40(4), 310–316.

Chan, F.T.S., Kumar, N. (2007). Global supplier development considering risk factors using fuzzy extended AHP-based approach. *OMEGA*, 35, 417–431.

Chang, J., de Búrca, C. (2016). An investigation into how small companies in London and the South East UK engage in IT offshore outsourcing and the impact of culture on this phenomenon. *Procedia Computer Science*, 100, 611–618. https://doi.org/10.1016/j.procs.2016.09.202.

Chauhan, P., Kumar, S., Sharma, R.K. (2015). Qualitative and quantitative approach to model offshore outsourcing barriers due to cultural differences. *International Journal of Strategic Business Alliances*, 4(2/3), 184. https://doi.org/10.1504/ijsba.2015.072034.

Dehdar, E., Azizi, A., Aghabeigi, S. (2018). Supply chain risk mitigation strategies in automotive industry: a review. 2018 IEEE International Conference on Industrial Engineering and Engineering Management (IEEM) (pp. 84–88). https://doi.org/10.1109/IEEM.2018.8607626.

Dhone, N.C., Kamble, S.S. (2015). Scale development for supply chain operational performance model in Indian automobile industry by using exploratory factor analysis. *International Journal of Logistics Systems and Management*, 22(2), 210–249.

Di Mauro, C., Fratocchi, L., Orzes, G., Sartor, M. (2018). Offshoring and backshoring: a multiple case study analysis. *Journal of Purchasing and Supply Management*, 24(2), 108–134. https://doi.org/10.1016/j.pursup.2017.07.003.

Dolgui, A., Proth, J.-M. (2013). Outsourcing: definitions and analysis. *International Journal of Production Research*, 51(23–24), 6769–6777. https://doi.org/10.1080/00207543.2013.855338.

Duhamel, F., Carbone, V., Moatti, V. (2016). The impact of internal and external collaboration on the performance of supply chain risk management. *International Journal of Logistics Systems and Management*, 23(4), 534–557.

Ecommerce News (2017). By Micah Maidenberg Oct. 25, 2017, Section B, Page 5 of the New York edition with the headline: E-Commerce Is Making Warehouses a Hot Property.

Ekici, A. (2013). An improved model for supplier selection under capacity constraint and multiple criteria. *International Journal of Production Economics*, 141(2), 574–581.

Ellram, L.M., Tate, W.L., Feitzinger, E.G. (2013). Factor-market rivalry and competition for supply chain resources. *Journal of Supply Chain Management*, 49(1), 29–46.

Errico, M., De Noni, I., Teodori, C. (2022). SMEs' financial risks in supply chain trade with large companies: the case of Italian automotive component industry. *Journal of General Management*, 47(2), 126–137.

Eydi, A., Fazli, L. (2019). A decision support system for single-period single sourcing problem in supply chain management. *Soft Computing*, 23, 13215–13233.

Fallahpour, A., Nayeri, S., Sheikhalishahi, M., Wong, K.Y., Tian, G., Fathollahi-Fard, A.M. (2021). A hyper-hybrid fuzzy decision-making framework for the sustainable-resilient supplier selection problem: a case study of Malaysian Palm oil industry. *Environmental Science and Pollution Research International*, 28, 1–21.

Ghadge, A., Weiß, M., Caldwell, N.D., Wilding, R. (2019). Managing cyber risk in supply chains: a review and research agenda. *Supply Chain Management: An International Journal*, 25(2), 223–240. https://doi.org/10.1108/scm-10-2018-0357.

Ghodsypour, S.H., O'Brien, C. (2001). The total cost of logistics in supplier selection, under conditions of multiple sourcing, multiple criteria and capacity constraint. *International Journal of Production Economics*, 73(1), 15–27.

Gray, J.V., Skowronski, K., Esenduran, G., Johnny Rungtusanatham, M. (2013). The reshoring phenomenon: what supply chain academics ought to know and should do. *Journal of Supply Chain Management*, 49(2), 27–33, 221.

Gunasekaran, A., Irani, Z., Choy, K.L., Filippi, L., Papadopoulos, T. (2015). Performance measures and metrics in outsourcing decisions: a review for research and applications. *International Journal of Production Economics*, 161, 153–166.

Gupta, T.K., Singh, V. (2015). A systematic approach to evaluate supply chain management environment index using graph theoretic approach. *International Journal of Logistics Systems and Management*, 21(1), 1–45.

Gurtu, A., Johny, J. (2021). Supply chain risk management: literature review. *Risks*, 9, 16. https://doi.org/10.3390/risks9010016.

Hahn, E.D., Bunyaratavej, K. (2010). Services cultural alignment in offshoring: the impact of cultural dimensions on offshoring location choices. *Journal of Operations Management*, 28(3), 186–193.

Hahn, E.D., Bunyaratavej, K., Doh, J.P. (2011). Impacts of risk and service type on nearshore and offshore investment location decisions. *Management International Review*, 51(3), 357–380.

Handley, S.M., Benton, W.C. (2012). The influence of exchange hazards and power on opportunism in outsourcing relationships. *Journal of Operations Management*, 30(1), 55–68.

Hansen, C., Mena, C., Aktas, E. (2018). The role of political risk in service offshoring entry mode decisions. *International Journal of Production Research*, 57, 1–17. https://doi.org/10.1080/00207543.2018.151860.

Hansen, C., Mena, C., Skipworth, H. (2017). Exploring political risk in offshoring engagements. *International Journal of Production Research*, 55(7), 2051–2067. https://doi.org/10.1080/00207543.2016.1268278.

Hughes, D.L., Rana, N.P., Dwivedi, Y.K. (2020). Elucidation of IS project success factors: an interpretive structural modelling approach. *Annals of Operations Research*, 285, 35–66. https://doi.org/10.1007/s10479-019-03146-w.

IBM Security. (2021). Cost of insider threats global report. Technical Report, Ponemon Institute.

Ishizaka, A., Bhattacharya, A., Gunasekaran, A., Dekkers, R., Pereira, V. (2019). Outsourcing and offshoring decision making. *International Journal of Production Research*, 57(13), 4187–4193. https://doi.org/10.1080/00207543.2019.160369.

Jain, V., Sangaiah, A.K., Sakhuja, S. (2018). Supplier selection using fuzzy AHP and TOPSIS: a case study in the Indian automotive industry. *Neural Computing and Applications*, 29, 555–564.

Jin, Y. (2000). Fuzzy modeling of high-dimensional systems: complexity reduction and interpretability improvement. *IEEE Transactions on Fuzzy Systems*, 8(2), 212–221. https://doi.org/10.1109/91.842154.

Johnson, D.M., Graman, G.A. (2015). Outsourcing practices of Midwest US public universities. *International Journal of Business Excellence*, 8(3), 268–297.

Jüttner, U., Peck, H., Christopher, M. (2003). Supply chain risk management: outlining an agenda for future research. *International Journal of Logistics Research and Applications*, 6(4), 197–210.

Kalyan Singhal, Jaya Singhal (2012). *Opportunities for developing the science of operations and supply-chain management*, 30(3), 245–252. doi:10.1016/j.jom.2011.11.002.

Kar, A.K., Pani, A.K. (2014). Exploring the importance of different supplier selection criteria. *Management Research Review*, 37(1), 89–105.

Karlsen, J.T., Sæther, H.S., Oorschot, K.E.V., Vaagaasar, A.L. (2021). Managing trust and control when offshoring information systems development projects by adjusting project goals. *International Journal of Technology Management*, 85(1), 42–77.

Katsaliaki, K., Galetsi, P., Kumar, S. (2022). Supply chain disruptions and resilience: a major review and future research agenda. *Annals of Operations Research*, 319, 965–1002.

Kaur, H., Singh, S.P., Majumdar, A. (2019). Modelling joint outsourcing and offshoring decisions. *International Journal of Production*

König, A., Spinler, S. (2016). The effect of logistics outsourcing on the supply chain vulnerability of shippers: Development of a conceptual risk management framework. *The International Journal of Logistics Management*, 27(1), 122–141. https://doi.org/10.1108/IJLM-03-2014-0043.

KPMG. (2017). Available at: https://home.kpmg/uk/en/home/services/advisory/risk-consulting/technology-risk/cyber-security.html [accessed 11 Sep. 2022].

Kumar, A., Mangla, S.K., Luthra, S., Ishizaka, A. (2019). Evaluating the human resource related soft dimensions in green supply chain management implementation. *Production Planning & Control*, 30, 1–17. https://doi.org/10.1080/09537287.2018.1555342.

Kumar, S., Kumar, R., Chauhan, P. (2014). ISM approach to model offshore outsourcing risks. *International Journal of Production Management and Engineering*, 2(2014), 101–111.

Kumar Sharma, S., Bhat, A. (2014). Supply chain risk management dimensions in Indian automobile industry: a cluster analysis approach. *Benchmarking: An International Journal*, 21(6), 1023–1040. https://doi.org/10.1108/BIJ-02-2013-0023.

Lakri, S., Dallery, Y., Jemai, Z. (2015). Measurement and management of supply chain performance: practices in today's large companies. *Supply Chain Forum: An International Journal*, 16, 16–30.

Langley, J. and Capgemini (2012). The State of Logistics Outsourcing -2012 Third-Party Logistics: Results and Findings of the 16th Annual Study, Capgemini Consulting, USA.

Leimeister, S. (2010). *IT Outsourcing Governance: Client Types and Their Management Strategies*. Gabler, Wiesbaden.

Lin, N. (2020). Designing global sourcing strategy for cost savings and innovation: a configurational approach. *Management International Review*, Springer, 60(5), 723–753.

Lockamy III, A., McCormack, K. (2010). Analyzing risks in supply networks to facilitate outsourcing decisions. *International Journal of Production Research*, 48(2), 593–611.

Logistics & Supply Chain Management, Martin Christopher Financial Times Prentice Hall, 2011 - Business & Economics - 276 pages.

Luo, Y.D., Wang, S.L., Jayaraman, V., Zheng, Q.Q. (2013). Governing business process offshoring: properties, processes, and preferred modes. *Journal of World Business*, 48(3), 407–419.

Luthra, S., Garg, D., Haleem, A. (2014). Greening the supply chain using SAP-LAP analysis: a case study of an auto ancillary company in India. *International Journal of Business Excellence*, 7(6), 724–746.

Luthra, S., Luthra, S., Haleem, A. (2015). Hurdles in implementing sustainable supply chain management: an analysis of Indian automobile sector. *Procedia - Social and Behavioral Sciences*, 189, 175–183.

Mahdiraji, H.A., Kamardi, A.A., Beheshti, M., Razavi Hajiagha, S.H., Rocha-Lona, L. (2022). Analysing supply chain coordination mechanisms dealing with repurposing challenges during Covid-19 pandemic in an emerging economy: a multi-layer decision making approach. *Operations Management Research*, 15, 1341–1360.

Mehta, A., Mehta, N. (2017). Moving towards an integrated framework of IT-outsourcing success. *Journal of Global Information Technology Management*, 20(3), 171–194.

Mentzer, J.T., Flint, D.J., & Hult, G.T.M. (2001). Logistics Service Quality as a Segment-Customized Process. *Journal of Marketing*, 65(4), 82–104. https://doi.org/10.1509/jmkg.65.4.82.18390.

Mihalache, M., Mihalache, O.R. (2020). What is offshoring management capability and how do organizations develop it? A study of Dutch IT service providers. *Management International Review*, 60, 37–67. https://doi.org/10.1007/s11575-019-00407-5.

Munjal, S., Requejo, I., Kundu, S.K. (2018). Offshore outsourcing and firm performance: moderating effects of size, growth and slack resources. *Journal of Business Research*, 103, 484–494. https://doi.org/10.1016/j.jbusres.2018.01.014.

Najmi, A., Gholamian, M.R., Makui, A. (2013). Supply chain performance models: a literature review on approaches, techniques, and criteria. *Journal of Operations and Supply Chain Management*, 6, 94–113.

Naqvi, M.A., Amin, S.H. (2021). Supplier selection and order allocation: a literature review. *Journal of Data, Infrastructure and Management*, 3, 125–139.

Nordås, H.K. (2020). Make or buy: offshoring of services functions in manufacturing. *Review of Industrial Organization*, 57, 351–378. https://doi.org/10.1007/s11151-020-09771-1.

Pai, F.Y. (2015). How supplier exercised power affects the cooperative climate, trust and commitment in buyer-supplier relationships. *International Journal of Business Excellence*, 8(5), 662–673.

Patrucco, A.S., Scalera, V.G., Luzzini, D. (2016). Risks and governance modes in offshoring decisions: linking supply chain management and international business perspectives. *Supply Chain Forum*, 17(3), 170–182. https://doi.org/10.1080/16258312.2016.1219616.

Peck, H. (2006). Reconciling supply chain vulnerability, risk and supply chain management. *International Journal of Logistics Research and Applications*, 9(2), 127–142.

Paul D. Cousins, Robert Spekman, Strategic supply and the management of inter- and intra-organisational relationships, *Journal of Purchasing and Supply Management*, Volume 9, Issue 1, 2003, Pages 19–29, ISSN 1478–4092.

Pisani, N., Ricart, J.E. (2016). Offshoring of services: a review of the literature and organizing framework. *Management International Review*, 56(3), 385–424.

Rahman, H.U., Raza, M., Afsar, P., Khan, H.U., Nazir, S. (2020). Analyzing factors that influence offshore outsourcing decision of application maintenance. *IEEE Access*, 8, 183913–183926.

Raiborn, C., Butler, J., Massoud, M. (2009). Outsourcing support functions: Identifying and managing the good, the bad, and the ugly. *Business Horizons*, 52, 347–356. https://doi.org/10.1016/j.bushor.2009.02.005.

Rane, S.B., Kirkire, M.S. (2017). Interpretive structural modelling of risk sources in medical device development process. *International Journal of System Assurance Engineering and Management*, 8, 451–464. https://doi.org/10.1007/s13198-015-0399-6.

Rao, M.T. (2004). Key issues for global IT sourcing: country and individual factors. *Information Systems Management*, 21(3), 16–21.

Ray, B.K., Tao, S., Olkhovets, A., Subramanian, D. (2013). A decision analysis approach to financial risk management in strategic outsourcing contracts. *EURO Journal on Decision Processes*, 1, 187–203. https://doi.org/10.1007/s40070-013-0013-6.

Rezaei, A., Aghsami, A., Rabbani, M. Supplier selection and order allocation model with disruption and environmental risks in centralized supply chain. *International Journal of System Assurance Engineering and Management*, 12, 1036–1072. https://doi.org/10.1007/s13198-021-01164-1.

Salehi Heidari, S., Khanbabaei, M., Sabzehparvar, M. (2018). A model for supply chain risk management in the automotive industry using fuzzy analytic hierarchy process and fuzzy TOPSIS. *Benchmarking: An International Journal*, 25(9), 3831–3857. https://doi.org/10.1108/bij-11-2016-0167.

Sawik, T. (2013). Selection and protection of suppliers in a supply chain with disruption risks. *International Journal of Logistics Systems and Management*, 15(January), 143–159.

Sharma, P., Sangal, A.L. (2019). Building a hierarchical structure model of enablers that affect the software process improvement in software SMEs-a mixed-method approach. *Computer Standards & Interfaces*, 66, 103350.

Sharma, R.K. (2022). Examining interaction among supplier selection strategies in an outsourcing environment using ISM and fuzzy logic approach. *International Journal of System Assurance Engineering and Management*, 13, 2175–2194. https://doi.org/10.1007/s13198-022-01624-2.

She, A.H., Zarour, M., Alenezi, M., Sarkar, A.K., Agrawal, A., Kumar, R., Khan, R.A. (2020). Healthcare data breaches: insights and implications. *Healthcare (Basel)*, 8(2), 133. https://doi.org/10.3390/healthcare8020133. PMID: 32414183; PMCID: PMC7349636.

Sheel, A., Singh, Y.P., Nath, V. (2020). Managing agility in the downstream petroleum supply chain. *International Journal of Business Excellence*, Interscience Enterprises Ltd, 20(2), 269–294.

Sindhwani, R., Kumar, R., Behl, A., Singh, P.L., Kumar, A., Gupta, T. (2022). Modelling enablers of efficiency and sustainability of healthcare: A m-TISM approach. *Benchmarking: An International Journal*, 29(3), 767–792. https://doi.org/10.1108/BIJ-03-2021-0132.

Skowronski, K., Benton, W.C., Hill, J.A. (2020). Perceived supplier opportunism in outsourcing relationships in emerging economies. *Journal of Operations Management*, 66, 989–1023. https://doi.org/10.1002/joom.1123.

Smite, D., Calefato, F., Wohlin, C. (2015). Cost savings in global software engineering: where's the evidence? *IEEE Software*, 32(4), 26–32. https://doi.org/10.1109/ms.2015.102.

Søderberg, A.M., Krishna, S., Bjørn, P. (2013). Global software development: commitment, trust and cultural sensitivity in strategic partnerships. *Journal of International Management*, 19(4), 347–361.

Stringfellow, A., Teagarden, M.B., Nie, W. (2008). Invisible costs in offshoring services work. *Journal of Operations Management*, 26(2), 164–179.

Surange, V.G., Bokade, S.U. (2022). Ranking of Critical Risk Factors in the Indian Automotive Supply Chain Using TOPSIS with Entropy Weighted Criterions. In: Chaurasiya, P.K., Singh, A., Verma, T.N., Rajak, U. (eds) *Technology Innovation in Mechanical Engineering*. Lecture Notes in Mechanical Engineering. Springer, Singapore. https://doi.org/10.1007/978-981-16-7909-4_46.

Taherdoost, H., Brard, A. (2019). Analyzing the process of supplier selection criteria and methods. *Procedia Manufacturing*, 32(1), 1024–1034.

Tate, W.L., Ellram, L.M. (2012). Service supply management structure in offshore outsourcing. *Journal of Supply Chain Management*, 48(4), 8–29.

Urciuoli, L. (2015). Cyber-resilience: a strategic approach for supply chain management. *Technology Innovation Management Review*, 5(4), 13–18.

Urciuoli, L., Hintsa, J. (2016). Adapting supply chain management strategies to security -an analysis of existing gaps and recommendations for improvement. *International Journal of Logistics Research and Applications*, 20(3), 276–295.

Uygun, Y., Gotsadze, N., Schupp, F., Gzirishvili, L., Tindjou Nana, B.S. (2022). A holistic model for understanding the dynamics of outsourcing. *International Journal of Production Research*, 61, 1–31

Vanalle, R.M., Lucato, W.C., Ganga, G.M.D., Alves Filho, A.G. (2020). Risk management in the automotive supply chain: an exploratory study in Brazil. *International Journal of Production Research*, 58(3), 783–799. https://doi.org/10.1080/00207543.2019.1600762.

Wang, L., Foerstl, K., Zimmermann, F. (2017). Supply Chain Risk Management in the Automotive Industry: Cross-Functional and Multi-tier Perspectives. In: Abele, E., Boltze, M., Pfohl, H.C. (eds) *Dynamic and Seamless Integration of Production, Logistics and Traffic*. Springer, Cham. https://doi.org/10.1007/978-3-319-41097-5_7.

Wankhade, N., Kundu, G.K. (2020). Modelling the enablers to explore the driving power, dependence and strategic importance in achieving SC agility. *International Journal of Value Chain Management*, 11(1), 63–95.

Winkler, J.K., Dibbern, J., Heinzl, A. (2008). The impact of cultural differences in offshore outsourcing-case study results from German-Indian application development projects. *Information Systems Frontiers*, 10(2), 243–258.

Xue, L., Zhang, C., Ling, H., Zhao, X. (2013). Risk mitigation in supply chain digitization: system modularity and information technology governance. *Journal of Management Information Systems*, 30(1), 325–325.

Index

For Product Safety Concerns and Information please contact our EU
representative GPSR@taylorandfrancis.com
Taylor & Francis Verlag GmbH, Kaufingerstraße 24, 80331 München, Germany

www.ingramcontent.com/pod-product-compliance
Ingram Content Group UK Ltd.
Pitfield, Milton Keynes, MK11 3LW, UK
UKHW021828240425
457818UK00006B/121

* 9 7 8 1 0 3 2 4 6 0 5 7 4 *